ELEMENTS OF CIVIL ENGINEERING

ELEMENTS OF CIVIL ENGINEERING

(As per the Syllabus of Gujarat Technological University)

S.S. Bhavikatti
Emeritus Fellow (AICTE)
BVB College of Engineering and Technology, Hubli
(Formerly Principal, RYMEC, Bellary,
Professor & Dean
SDMCET, Dharwad and NITK, Surathkal)

PUBLISHING FOR ONE WORLD

NEW AGE INTERNATIONAL (P) LIMITED, PUBLISHERS
New Delhi • Bangalore • Chennai • Cochin • Guwahati • Hyderabad
Jalandhar • Kolkata • Lucknow • Mumbai • Ranchi
Visit us at www.newagepublishers.com

Published by New Age International (P) Ltd., Publishers
First Edition: 2009
Reprint: 2013

BRANCHES

- **Bangalore** 37/10, 8th Cross (Near Hanuman Temple), Azad Nagar, Chamarajpet, Bangalore-560 018
  Tel.: (080) 26756823, Telefax: 26756820, E-mail: bangalore@newagepublishers.com
- **Chennai** 26, Damodaran Street, T. Nagar, Chennai-600 017
  Tel.: (044) 24353401, Telefax: 24351463, E-mail: chennai@newagepublishers.com
- **Cochin** CC-39/1016, Carrier Station Road, Ernakulam South, Cochin-682 016
  Tel.: (0484) 2377004, Telefax: 4051303, E-mail: cochin@newagepublishers.com
- **Guwahati** Hemsen Complex, Mohd. Shah Road, Paltan Bazar, Near Starline Hotel, Guwahati-781 008
  Tel.: (0361) 2513881, Telefax: 2543669, E-mail: guwahati@newagepublishers.com
- **Hyderabad** 105, 1st Floor, Madhiray Kaveri Tower, 3-2-19, Azam Jahi Road, Nimboliadda, Hyderabad-500 027
  Tel.: (040) 24652456, Telefax: 24652457, E-mail: hyderabad@newagepublishers.com
- **Kolkata** RDB Chambers (Formerly Lotus Cinema) 106A, 1st Floor, S.N. Banerjee Road, Kolkata-700 014
  Tel.: (033) 22273773, Telefax: 22275247, E-mail: kolkata@newagepublishers.com
- **Lucknow** 16-A, Jopling Road, Lucknow-226 001
  Tel.: (0522) 2209578, 4045297, Telefax: 2204098, E-mail: lucknow@newagepublishers.com
- **Mumbai** 142C, Victor House, Ground Floor, N.M. Joshi Marg, Lower Parel, Mumbai-400 013
  Tel.: (022) 24927869, Telefax: 24915415, E-mail: mumbai@newagepublishers.com
- **New Delhi** 22, Golden House, Daryaganj, New Delhi-110 002
  Tel.: (011) 23262368, 23262370, Telefax: 43551305, E-mail: sales@newagepublishers.com

ISBN : 978-81-224-2682-3

₹ 150.00

C-12-08-6521

Printed in India at Print-O-Pack, Delhi.
Typeset at Goswami Associates, Delhi.

PUBLISHING FOR ONE WORLD

NEW AGE INTERNATIONAL (P) LIMITED, PUBLISHERS
7/30 A, Darya Ganj, New Delhi-110002
Visit us at **www.newagepublishers.com**

Preface

Gujarat Technological University has introduced the course Elements of Civil Engineering for the students of civil engineering discipline. It covers almost all basic subjects of civil engineering such as surveying, materials of construction, construction technology, hydrology, transportation engineering and traffic engineering. It is difficult for a student to look for the required materials from specialized books in these subjects of civil engineering. Hence there is need for a book specially for the course.

The author, from his experience in writing similar book for Visweswaraya Technological University has written this book to make it more student friendly, using simple language, illustrating with neat figures and solving relevant problems in proper manner. It is hoped that students and teaching staff of Gujarat Technological University will accept it widely. Suggestions for improvements are most welcome from the readers.

S.S. Bhavikatti

Syllabus

I Scope of Civil Engineering

Introduction: Impact of Infrastructural Development on the Economy of a Country, Role of Civil Engineers, Importance of Planning, Scheduling and Construction Management.

II Surveying

Introduction: Surveying and levelling, Object and uses, Primary divisions, Fundamental principles, Classification of surveying, Plans and maps, Scales, Units of measure.

Linear measurements: Methods, Instruments used in chaining; Chain surveying, Ranging; Obstacles; Errors in chaining, Tape corrections, Conventional symbols.

Compass surveying: Types and uses of compass, Bearings, Whole Circle Bearings, and Reduced Bearings, Computation of angles; Meridians; declinations and dip of needle; Local attraction; Chain and compass surveying field work.

Elevation measurements: Levelling, object and uses, terms used in levelling, levelling instruments, methods of levelling, recording and methods of reducing, errors in levelling, contours; characteristics and applications.

Areas and volumes: Use of a Planimeter.

Modern Tools of Surveying and Mapping: Introduction to Theodolite, Electronic Distance Measurement Instruments, Total Station, Global Positioning System, Remote Sensing and Geographic Information System.

III Construction Materials

Requirement, types, uses, properties and importance of Civil Engineering materials like Stone, Bricks, Lime, Cement, Ferrous and Non-ferrous Metals, Ceramic Materials, Timber, Sand, Aggregate, Mortar and Concrete, Paints and Varnishes, Glass, Plastic, Conducting, Magnetic and Miscellaneous Materials.

IV Elements of Building Construction

Planning: Elementary principles and basic requirements of a building planning, layout of residential and industrial buildings.

Construction: Classification of buildings based upon occupancy and structure, Design Loads, Common building components, their functions, and nominal dimensions. Elements of building drawing. Introduction to building byelaws.

V Water Resources Development

Elementary Hydrology, Sources of water, Watershed Development, Water requirements and its conservation, Hydraulic Structures of Storage, Water Conveyance System: Canals; Water Conduits.

VI Transportation Engineering

Role of Transportation in National development, Transportation Ways, Surface Transportation and Aviation, Elements of Highway materials properties and highway construction, BOT Projects for Highways, Elements of Traffic Engineering and Traffic Control.

Contents

1

Scope of Civil Engineering

Civil engineering is the oldest branch of engineering which is growing right from the stone age of civilisation. American society of civil engineering defines *civil engineering as the profession in which a knowledge of the mathematical and physical sciences gained by study, experience and practice is applied with judgement to develop ways to utilise economically the materials and forces of the nature for the progressive well being of man.*

In this chapter various fields of civil engineering are listed and their scope is explained. Impact of infrastructural development on the economy of a country, the role of civil engineers in it is explained. Finally the importance of planning scheduling and construction management in civil engineering is emphasised.

1.1 FIELDS OF CIVIL ENGINEERING AND THEIR SCOPE

Civil engineering may be divided into the following fields:

(*i*) Building materials
(*ii*) Building construction
(*iii*) Structural engineering
(*iv*) Geotechnical engineering
(*v*) Hydraulics, water resources and irrigation engineering
(*vi*) Water supply and sanitary engineering
(*vii*) Environmental engineering
(*viii*) Transportation engineering
(*ix*) Town planning and architecture
(*x*) Surveying
(*xi*) Drawing
(*xii*) Estimation and specification
(*xiii*) Management techniques
(*xiv*) Computer application.

(*i*) **Building Materials.** Shelter is the basic need of civilised society. Stones, bricks, timber and lime concrete are the traditional materials used for the construction of houses and other buildings. The invention of cement and concrete has prodivided durable buildings. Reinforced concrete which is composite construction of steel and concrete has helped in building large structurer. Steel, aluminium, glass, plastics, glazed tiles, plaster of paris, linoleum, paints and varnishes have improved the quality of buildings. Improved versions of many building materials keep on appearing in the market regularly. A civil engineer has to make use of all these materials judiciously.

(*ii*) **Building Construction.** Fast rate of urbanisation and increase in the cost of land has forced to go for the vertical growth of buildings. In metropolitan cities, 25 storey buildings are becoming common. Even in small towns 3 to 4 storey buildings have become common. This requirement has brought in new building technologies. Continuous research and development in construction technology is going on to see that huge investment in building is utilised at the earliest. Civil engineers have to look into the problems of rural areas and urban poor people also. There are 23 million Indians without home. Low cost housing is the need of the day to make poor people afford their own houses.

(*iii*) **Structural Engineering.** Before building a structure, it should be analysed and designed to decide about its size to resist the possible forces coming on it. The structure should be safe and at the same time its components should be as small as possible. Up to mid sixties lot of improvements were seen in the classical methods of analysis. Need of tall structures and improvements in computers gave rise to matrix method and finite element method of analysis. Requirement of large column free structures gave rise to analysis and design of shell roofs (curved surfaces), geodetic towers and tension structures. Disasters due to earthquakes have made civil engineers to study earthquake forces and build earthquake resistant structures. It needs the knowledge of structural dynamics. A civil engineer has to not only give a safe structure but he has to give an economical structure also. Hence, there is need for studying mathematical optimisation techniques. All these aspects of analysis and design fall under structural engineering field.

(*iv*) **Geotechnical Engineering.** All structures have to finally transfer the load acting on them to soil safely. Soil property changes from place to place. Even in the same place it may not be uniform at different depth and in different seasons. Hence, a civil engineer has to properly investigate soil and decide about the safe load that can be spread on the soil. This branch of study in civil engineering is known as geotechnical engineering. Apart from finding safe bearing capacity for foundation of buildings, geotechnical engineering involves various studies required for the design of pavements, tunnels, earthen dam, canals and earth retaining structures. It involves study of ground inprovement techniques also.

(*v*) **Hydraulics, Water Resources and Irrigation Engineering.** Water is an important need for all living beings. Study of mechanics of water and its flow characteristics is another important field in civil engineering and it is known as hydraulics.

Requirement of water in cities for domestic purpose and for industries is continuously increasing. Rural areas need water for agricultural field also. Hence civil engineers have to look for new water resources and for storing them. This branch of civil engineering is known as water resources engineering.

Water stored in reservoirs by building bunds and dams should be brought to agricultural fields through canals and distributories. Study connected with this aspect is known as irrigation engineering.

(*vi*) **Water Supply and Sanitary Engineering.** When water is required for drinking purpose it should be purified and made potable. Purification of water and the technology involved in taking it to the houses is known as water supply engineering.

Waste waster and solid waste should be treated and disposed so that they do not create health hazard. This branch of civil engineering is known as sanitary engineering.

(*vii*) **Environmental Engineering.** Apart from tackling solid and waste water disposal civil engineers have to tackle air pollution problem also. Due to industrialisation air pollution is becoming a major problem. It is estimated that for every tonne of cement produced one tonne of CO_2 is released to environment. Vehicles also produce lot of CO_2. During the last one century, the environmental pollution has resulted in global warming by 4°C. An environmental disaster will be unavoidable if China, India and other developing countries start consuming as much energy and materials as the West did it in its march to industrialisation. Hence environmental engineering is emerging as an important field of study in civil engineering.

(*viii*) **Transportation Engineering.** Transportation facility is another important need. Providing good and economical road links is an important duty of civil engineers. It involves design and construction of base courses, suitable, surface finishes, cross drainage works, intersections, culverts, bridges and tunnels etc. Railways is another important long-way transport facility. Design, construction and maintainance of railway lines are parts of transportation engineering. Globalisation has resulted into requirement of airports and harbours. For proper planning of these transport facilities, traffic survey is to be carried out. Carrying out traffic survey and then planning, designing, construction and maintainance of roads, railways, bridges, tunnels, airports and harbours is known as transportation engineering.

(*ix*) **Town Planning and Architecture.** With the growth of population and industries new towns are coming up and existing ones are growing. Proper town planning is to be made by civil engineers. Structures should be aesthetically good also. Architecture covers this area. This field of civil engineering has grown up so much that it has become a separate branch of engineeing.

(*x*) **Surveying.** For planning all developmental activities, proper maps are required. The science of map making is known as surveying. Survey maps provide the relative positions of various objects of the area in the horizontal as well as vertical directions. Earlier conventional instruments like chain, tape, compasses, theodolites and levels were used for various measurements in surveying. In this electronic era the modern equipments like electronic distance meters and total stations are used for measurements. Modern technology like remote sensing has made surveying vast area in a short period possible.

(*xi*) **Drawing.** Drawing is the language of engineers. The survey maps and plans, the building description etc. are to be provided with neat scaled drawings.

(*xii*) **Estimation and Specification.** Civil engineers have to prepare estimation and detailed specifications for each and every work to be taken up.

(*xiii*) **Management Techniques.** Civil engineers must manage, men, materials and equipments efficiently. Since huge funds are to be handled in civil engineering projects, a civil engineer must know the basics in financial management and leagal obligations. Knowledge of management techniques is an asset to practising civil engineer.

(*xiv*) **Computer Applications.** Since the magnitude of designing the structures and storing information is increasing very fast nowadays civil engineers go for computer applications. Nowadays neat drawings are also produced using computers. There are a good number of civil engineering software commercially available.

1.2 IMPACT OF INFRASTRUCTURAL DEVELOPMENT ON THE ECONOMY OF THE COUNTRY

Civil engineering activities in the infrastructural development are:

(*i*) Good planning of towns and extension areas in the cities. Each extension area should be self sufficient in accommodating offices, educational institutions, markets, hospitals, recreational facilities and residential accommodation.

(*ii*) Assured water supply.

(*iii*) A good drainage system.

(*iv*) Pollution free environmental conditions.

(*v*) A well planned and built network of roads and road crossings.

(*vi*) Railways connections to all important cities and towns.

(*vii*) Airports and harbours of national and international standards.

Infrastructure also involves electricity supply, without assured electric supply no city town can develop. Internet and telephones are also desirable features.

Educational facility also forms part of infrastructure. Proximity of good primary and secondary schools to residential areas is desirable. Collegiate and professional education also form part of infrastructure of a city.

Good health care facility is a necessity. Primary health centres, specialised hospitals and doctors add to the desirable infrastructure facility.

Effect of infrastructure facilities are:

1. Connecting producing centres to marketing places minimise exploitation of producers by middlemen. Imports and exports became easy and as a result of which whole world becomes a village.
2. Improved irrigation facility enhances agricultural products and hence producers as well as consumers are benefitted.
3. Infrastructural facility develops scope for a number of industries and it creates job opportunities.
4. Improved éducation and health care give rise to skilled and healthy work force. Quality of life of the people is improved.
5. Utilisation of manpower for the benefit of mankind brings down antisocial activities.
6. In case of natural calamities assistance can be easily extended to the affected areas and misery of affected people minimised.
7. Infrastructural facility imporves defence system and peace exists in the country.
8. Improved economical power of the country brings a respectable status in the world.

The world has realised that a government should not involve itself in production and distribution but should develop infrastructure to create an atmosphere for economical development.

1.3 ROLE OF CIVIL ENGINEERS

A civil engineer has to conceive, plan, estimate, get approval, create and maintain all civil engineering activities. Civil engineer has very important role in the development of the following infrastructure:

(*i*) Measure and map the earth's surface.

(*ii*) Plan new townships and extension of existing towns.

(*iii*) Build the suitable structures for the rural and urban areas for various utilities.

(*iv*) Build tanks and dams to exploit water resources.

(*v*) Build river navigation and flood control projects.

(*vi*) Build canals and distributaries to take water to agricultural fields.

(*vii*) Purify and supply water to the needy areas like houses, schools, offices etc.

(*viii*) Provide and maintain communication systems like roads, railways, harbours and airports.

(*ix*) Devise systems for control and efficient flow of traffic.

(*x*) Provide and maintain solid and waste water disposal system.

(*xi*) Monitor land, water and air pollution and take measures to control them. Fast growing industrialisation has put heavy responsibilities on civil engineers to preserve and protect environment.

1.4 IMPORTANCE OF PLANNING, SCHEDULING AND CONSTRUCTION MANAGEMENT

Planning, Scheduling and Management form the basis for any construction activity. Though discussion here is about construction, the same principles hold good for any organisation.

Planning

Planning is the decision-making process about What, Where, Who and How (WWWH) to start a project.

What: An individual or a group of enterprising people plan to start a project. They form a group of high level managers. They identify goals, frame the objective and identify opportunities. The planning done by them may be called as strategic planning.

Where: The strategic planners decide the lime and the place of starting the organisation.

Who: The strategic planners identify the middle level managers and operational level planners to carry out the task. The middle level management deals with financial management and coordinates with operational planners and strategic planners.

How: The operational planners have to work at minute details of the work assigned, find the requirement of machinery and work force and plan day to day activities. They should be ready with alternative plans, if uncertainity creeps in at any stage.

Importance of Planning

Although events can never be predicted exactly, planning is important. Factors beyond control may interfare and upset plans, but no work can be achieved successfully without planning. Advantages of planning are:

(*i*) *It minimises uncertainities.* Certain uncertainities like availability of work force, equipment, raw materials etc., due to strikes and natural calamities may upset plans. However with short term alternate plans, many adverse effects of these uncertainties can be minimised. Without a plan, many uncertainties will creep into project work and hamper the other wings of work force. Hence a planning is a must.

(*ii*) *It exercises good control.* With proper planning there is optimum use of materials and man power, which result is cost reduction in construction industry. There will be good control on every wing of the construction activity. Cashflow for the activity is streamlined avoiding unnecessary holding of the funds or hampering of the work due to cash shortage. Hence planning is necessary for exercising good control on the construction work.

(*iii*) *Planning focusses on objectives.* A construction industry may have an objective of specializing itself in irrigation projects, road projects, building construction, producing precost products etc. It has equipment and expertise in achieving particular objectively. Planning helps the managers of all department to work for achieving the objective, periodically revise the plan in the interest of achieving the objective.

(*iv*) *Planning leads to success.* Occasionally unforeseen factors may affect success to some extent but planning puts the organisation definitely on the path to success. If the work is carried out as per the plan without any hinderance, an organisation can make good profit and become competent for obtaining more works.

Scheduling

A construction work may be proposed to be started on a particular day and completed within the specified period. The entire project is to be subdivided into a number of events. For example, the construction of a building may be divided into the following stages:

1. Mobilising
2. Laying Foundation
3. Building Super Structure
4. Finishing

Laying of foundation may be subdivided into the following events:

(*i*) Digging

(*ii*) Laying bed concrete

(*iii*) Building stone masonry courses

(*iv*) Providing coping concrete

(*v*) Pointing joints and

(*vi*) Filling the trenches.

Like this all stages involve a number of events. For each event, scheduled dates of completion are to be fixed and to complete each event in the prescribed days material and labour requirements are to be identified. For this, the tool used in construction industry is bar chart method, in which each activity with number of days required for completion are marked. Starting and completing day of works are shown. Network representation of the events is a superior method. In this method interconnectivitiy of various events are clearly indicated. It helps in identifying critical activities. Delay in starting and completing a critical activity delays the entire project. The path joining the critical activity is called critical path. *Identifying various events, time, labour, material requirements for each event and representing the sequence of events by bar charts or by network is called scheduling.*

Importance of Scheduling

Scheduling fixes up the precise task and responsibilities for all the personnel involved in the construction. The various persons involved are the owner, financial managers, architect, design engineer, site engineer, labour contractor, contractors for material supply etc. Without scheduling, coordination among the above persons cannot be achieved. Scheduling is necessary for the success of project implementation.

Construction Management

Management is the science and art of planning, organising, staffing, scheduling and executing the work. Planning, scheduling and executing work need well trained engineers. Achieving coordination among various people involved in the work needs proper management training. In fact in any work involving a common goal, management is essential to *coordinate* the individual efforts. It is necessary to maintain a proper work environment for all the individuals to work together. A manager not only *directs* his subordinates but also *motivates* them to do the work. *Decision-making and leadership* are the basic qualities of a good manager. Good *communication skill* is to be developed by the managers with their seniors and subordinates to keep the work going on smoothly and review the execution of various activities.

Importance of Construction Management

(*i*) Planning and Scheduling helps in recruiting right people at right time.

(*ii*) Leadership, decision-making ability and properly directing subordinates avoid the wastage of labour.

(*iii*) Communication skill and proper coordination helps in maintaining the continuity of the works and avoids delays.

(*iv*) Motivating the work force with various schemes and awards help in extracting maximum work and increases efficiency.

(*v*) Proper management avoids locking of funds and ensures proper inflow of cash. As a result.

(*a*) Resource utilisation is optimised

(*b*) Efficiency is achieved

(*c*) Construction cost is reduced and

(*d*) Schedule is maintained.

The organisation achieves good name and makes profit. The success of the company benefits itself and the society.

QUESTIONS

1. Briefly explain the scope of civil engineering.
2. What are the civil engineering infrastructure systems required for the socio-economical development of a country ?
3. Write short note on impact of infrastructure development on the economy of the country.
4. Explain briefly the role of a civil engineer.
5. Why planning is important in the construction works ?
6. What do you understand by planning of a work ?
7. What is scheduling? Explain its importance in the construction work.

2

Introduction to Surveying

Surveying is *the art of making measurements of objects on, above or beneath the ground to show their relative positions on paper.* The relative position required is either horizontal, or vertical, or both.

Less precisely the term **Surveying** is used to the *measurement of objects in their horizontal positions. Measurements to deteremine their relative vertical positions is known as* **levelling**.

2.1 OBJECT AND USES OF SURVEYING

As stated in the definition, object of surveying is to show relative positions of various objects of an area on paper and produce plan or map of that area. Various uses of surveying are listed below:

(*i*) Plans prepared to record property lines of private, public and government lands help in avoiding unnecessary controversies.

(*ii*) Maps prepared for marking boundaries of countries, states, districts etc., avoid disputes.

(*iii*) Locality plans help in identifying location of houses and offices in the area.

(*iv*) Road maps help travellers and tourist.

(*v*) Topographic maps showing natural features like rivers, streams, hills, forests help in planning irrigation projects and flood control measures.

(*vi*) For planning and estimating project works like roads, bridges, railways, airports, water supply and waste water disposal surveying is required.

(*vii*) Marine and hydrographic survey helps in planning navigation routes and harbours.

(*viii*) Military survey is required for strategic planning.

(*ix*) Mine surveys are required for exploring minearl wealth.

(*x*) Geological surveys are necessary for determining different strata in the earth crust so that proper location is found for reservoirs.

(*xi*) Archeological surveys are useful for unearthing relics of antiquity.

(*xii*) Astronomical survey helps in the study of movements of plants and for calculating local and standard times.

2.2 PRIMARY DIVISIONS IN SURVEYING

The earth is an oblate spheroid, length of equatorial axis being 12756.75 km and polar axis being 12713.80 km. Since the difference between these two axes and irregularities on the earth surface are very small (Note. Height of Mount Everest is 8.79 km) compared to these two axes, the earth may be treated as a sphere, Figure 2.1 shows a circular plane passing through a point *A* on the earth surface. The gravitational force is always directed towards the centre of the earth.

Hence, the plumb-line shown in Fig. 2.1. is a *vertical line*. Line perpendicular to vertical line (tangential to earth surface) is known as *horizontal line*. In surveying all measurement at any point are in the direction of these two lines.

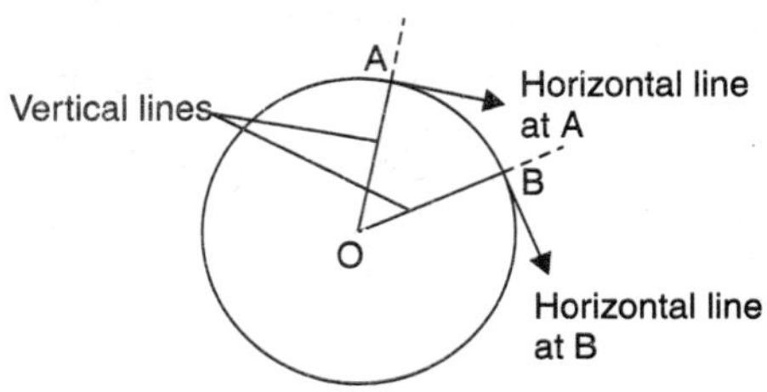

Fig. 2.1 Vertical and horizontal lines

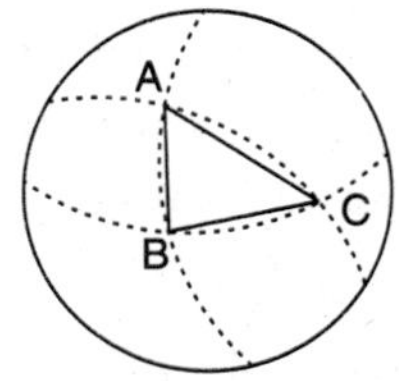

Fig. 2.2 Plane and spherical triangles

Obviously, the vertical and horizontal lines at another point *B* are not parallel to the respective lines at *A*. It should be noted that all lines lying on the earth's surface are curved lines and all triangles are spherical triangles as shown in Fig. 2.2. Hence, surveying involves spherical trigonometry.

If the area to be surveyed is small, the curvature of the earth may be neglected and all plumb lines treated as the same vertical. Hence, the lines normal to plumb line at any point in the area are treated as the same horizontal. All triangles in the area may be treated as plane triangles.

The survey in which earth curvature is neglected is called *Plane Surveying* and the survey in which earth's curvature is considered is known as *Geodetic Surveying*.

No definite value can be assigned to the area up to which a survey may be treated as plane, since the degree of accuracy required forms the controlling factor. However, the following points should be noted:

(*i*) The length of an arc of 1.2 km on earth's mean surface is only 1 mm more than the straight line connecting those two points.

(*ii*) The sum of the interior angles of a geometrical figure laid on the surface of the earth differs from that of the corresponding figure only to the extent of one second for about 200 square kilometres of area.

Hence, in most of engineering projects plane surveying is used. The geodetic surveying is used to determine the precise positions of control stations on the surface of the earth to which plane survey details are connected in works of larger magnitude like preparing maps of countries. Thus, in surveying there are two primary divisions *viz. Geodetic Surveying* and *Plane Surveying*.

2.3 FUNDAMENTAL PRINCIPLES OF SURVEYING

To get accurate results in surveying one should follow the following fundamental principles is surveying:

(*i*) Work from whole to part

(*ii*) Extra care in fixing new control points.

2.3.1 Work From Whole to Part

In surveying large areas, a system of control points are identified and they are located with high precision. Then secondary control points are located using lesser precise methods. The details of the localised areas are measured and plotted with respect to the secondary control points. This is called working from whole to part. This principle in surveying helps in localising the errors. If the surveying is carried out by adding localised areas errors accumulated and may become unacceptable when large area is covered.

2.3.2 Extra Care in Fixing New Control Points

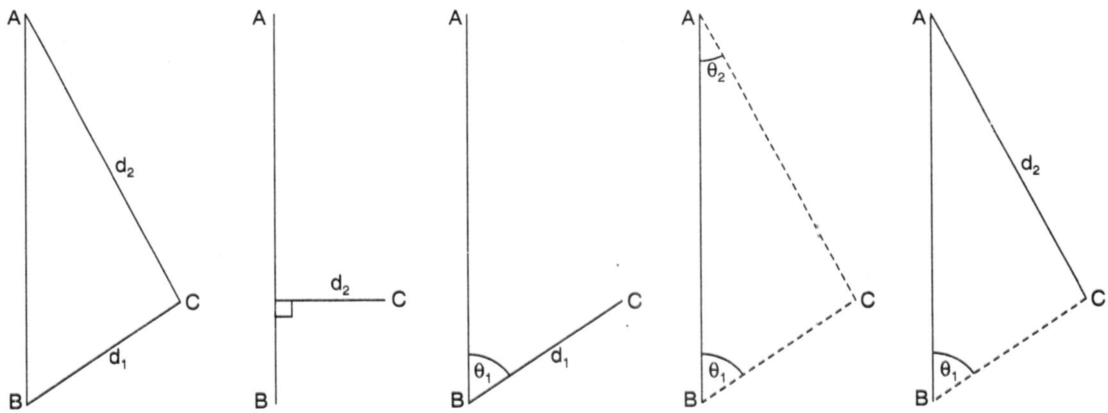

Fig. 2.3 Locating point C w.r.t. points A and B

Figure 2.3 shows the various methods of fixing point *C* with respect to already fixed points *A* and *B* by measuring sides, angles or setting perpendiculars. For fixing new control points (stations) with respect to already fixed points at least two independent process should be followed. If *A* and *B* are already located control points and with respect to them new control point *C* is to be located, apart from the minimum measurements required as shown in Fig. 2.3, one more measurement should be taken. Measuring the lengths of check lines and tie lines will also serve this purpose (Ref. Fig. 2.4).

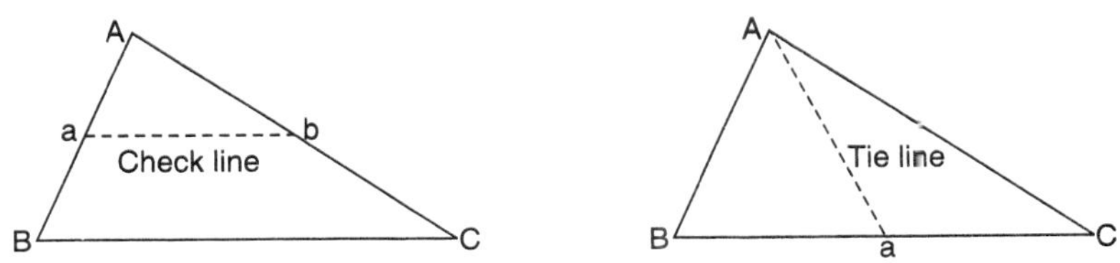

Fig. 2.4 Check line and tie line

2.4 CLASSIFICATION OF SURVEYING

Surveying may be classified on the following basis:

(*i*) Nature of the survey field

(*ii*) Object of survey

(*iii*) Instruments used and

(*iv*) The methods employed.

2.4.1 Classification Based on Nature of Survey Field

On this basis survey may be classified as land survey, marine or hydraulic survey and astronomical survey.

Land Survey. It involves measurement of various objects on land. This type of survey may be further classified as given below:

(*a*) *Topographic Survey.* It is meant for plotting natural features like rivers, lakes, forests and hills as well as man made features like roads, railways, towns, villages and canals.

(*b*) *Cadestal Survey.* It is for marking the boundaries of municipalities, villages, talukas, districts, states etc. The survey made to mark properties of individuals also come under this category.

(*c*) *City Survey.* The survey made in connection the construction of streets, water supply and sewage lines fall under this category.

Marine or Hydrographic Survey. Survey conducted to find depth of water at various points in bodies of water like sea, river and lakes fall under this category. Finding depth of water at specified points is known as sounding.

Astronomical Survey. Observations made to heavenly bodies like sun, stars etc., to locate absolute positions of points on the earth and for the purpose of calculating local time is known as astronomical survey.

2.4.2 Classification Based on Object of Survey

On the basis of object of survey the classification can be as engineering survey, military survey, mines survey, geological survey and archeological survey.

(*a*) *Engineering Survey:* The objective of this type of survey is to collect data for designing civil engineering projects like roads, railways, irrigation, water supply and sewage disposals. These surveys are further sub-divided into:

Reconnaissance Survey for determining feasibility and estimation of the scheme.

Preliminary Survey for collecting more information to estimate the cost of the project, and

Location Survey to set the work on the ground.

(*b*) *Military Survey:* This survey is meant for working out plans of strategic importance.

(*c*) *Mine Survey:* This is used for exploring mineral wealth.

(*d*) *Geological Survey:* This survey is for finding different strata in the earth's crust.

(*e*) *Archaeological Survey:* This survey is for unearthing relics of antiquity.

2.4.3 Classification Based on Instruments Used

Based on the instruments used, surveying may be classified as:

(*i*) Chain survey

(*ii*) Compass survey

(*iii*) Plane table survey

(*iv*) Theodolite survey

(*v*) Tacheometric survey

(*vi*) Modern survey using electronic distance meters and total station

(*vii*) Photographic and Aerial survey

The survey is taught to students mainly based on this classification.

2.4.4 Classification Based on Methods Employed

On this basis surveying is classified as triangulation and traversing.

(*i*) *Triangulation:* In this method control points are established through a network of triangles.

(*ii*) *Traversing:* In this scheme of establishing control points consists of a series of connected points established through linear and angular measurements. If the last line meets the starting point it is called as closed traverse. If it does not meet, it is known as open traverse [Ref. Fig. 2.5].

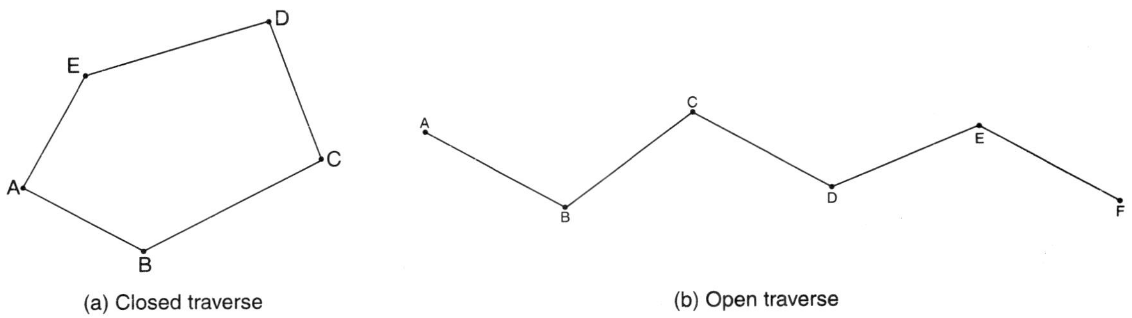

Fig. 2.5 Traversing

2.5 PLANS AND MAPS

As stated in the definition of surveying the objective of measurements is to show relative positions of various objects on paper. Such representations on paper is called plan or map. *A plan may be defined as the graphical representation of the features on, near or below the surface of the earth as projected on a horizontal plane to a suitable scale.*

However, since the surface of the earth is curved and that of the paper is plane, no part of the earth can be represented on such maps without distortion. If the area to be represented is small, the distortion is less and large scale can be used. Such representations are called plans. If the area to be represented is large, small, scales are to be used and distortion is large. Representation of larger areas are called maps. Representation of a particular locality in a municipal area is a plan while representation of a state/country is a map. There is no exact demarcation between a plan and map.

2.6 SCALES

It is not possible and also not desirable to make maps to one to one scale. While making maps all distances are reduced by a fixed proportion. That fixed proportion is called scale of the map. Thus, if 1 mm on the paper represents 1 metre on the ground, then the scale of the map is 1 mm = 1 m or 1 mm = 1000 mm or 1 : 1000. To make scale independent of the units it is preferable to use representative factor which may be defined as the ratio of one unit on paper to the number of units it represent on the ground. Thus 1 mm = 1 m is equivalent to

$$\text{RF} = \frac{1}{1000}$$

Apart from writing scale on map, it is desirable to show it graphically on it. The reason is, over the time, the paper may shrink and the scaling down the distances from map may mislead. The graphical scale should be sufficiently long (180 mm to 270 mm) and the main scale divisions should represent one, ten or hundred units so that it can be easily read.

The scale of a map is considered as

(*i*) large if it is greater than 1 cm = 10 m *i.e.*,

$$\text{RF} > \frac{1}{1000}$$

(*ii*) intermediate if it is between

$$\text{RF} = \frac{1}{1000} \quad \text{and} \quad \frac{1}{10{,}000}$$

(*iii*) small if $\text{RF} < \dfrac{1}{10{,}000}$

In general, scale selected should be as large as possible, since it is not possible for human eye to distinguish between two point if distance between them is less than 0.25 mm. The recommended scales for various types of surveys are as shown in Table 2.1.

2.7 TYPES OF GRAPHICAL SCALES

The following two types of scales are used in surveying:

(*i*) Plain Scale

(*ii*) Diagonal Scale.

2.7.1 Plain Scale

On a plain scale it is possible to read two dimensions directly such as unit and tenths. This scale is not drawn like ordinary foot rule (30 cm scale). If a scale of 1 : 40 is to be drawn, the markings are not like 4 m, 8 m, 12 m etc. at every 1 cm distance. Construction of such a scale is illustrated with the example given below:

■ **Example 2.1:** *Construct a plain scale of RF =* $\frac{1}{500}$ *and indicate 66 m on it.*

Solution. If the total length of the scale is selected as 20 cm, it represents a total length of 500 × 20 = 10000 cm = 100 m. Hence, draw a line of 20 cm and divide it into 10 equal parts.

Hence, each part correspond to 10 m on the ground. First part on extreme left is subdivided into 10 parts, each subdivision representing 1 m on the field. Then they are numbered as 1 to 10 from left to right as shown in Fig. 2.6. If a distance on the ground is between 60 and 70 m, it is picked up with a divider by placing one leg on 60 m marking and the other leg on subdivision in the first part. Thus field distance is easily converted to map distance.

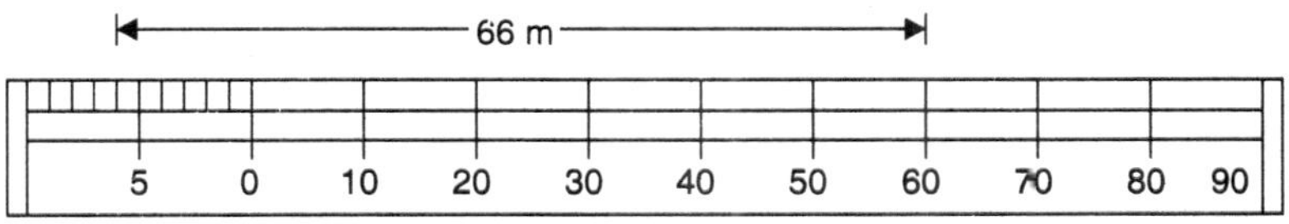

Fig. 2.6

IS 1491—1959 recommends requirements of metric plain scales designated as A, B, C, D, E and F as shown in Table 2.2. Such scales are commonly available in the market. They are made of either varnished cardboard or of plastic materials. Such scales are commonly used by surveyors and architects.

2.7.2 Diagonal Scale

In plain scale only unit and tenths can be shown whereas in diagonal scales it is possible to show units, tenths and hundredths. Units and tenths are shown in the same manner as in plain scale. To show hundredths, principle of similar triangle is used. If *AB* is a small length and its tenths are to be shown, it can be shown as explained with Fig. 2.7 below.

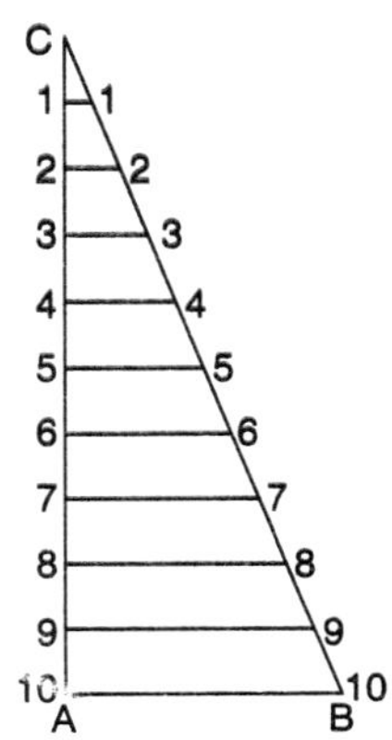

Fig. 2.7

Draw the line *AC* of convenient length at right angles to plain scale *AB*. Divide it into 10 equal parts. Join *BC*. From each tenth point on line *AC* draw lines parallel to *AB* till they meet line *BC*. Then line 1–1 represent $\frac{1}{10}$ th of *AB*, 6–6 represent $\frac{6}{10}$ th of *AB* and so on. Figure 2.8 shows the construction of diagonal scale with RF = $\frac{1}{500}$ and indicates 63.6 m.

Table 2.1 Recommended scales for various types of surveys

Type of survey	*Scale*	*RF*
1. Building sites	1 cm = 10 m or less (1 : 1000 or less)	$\frac{1}{1000}$ or less
2. Town planning schemes and reservoirs	1 cm = 50 m to 100 m (1 : 5000 to 1 : 10000)	$\frac{1}{5000}$ to $\frac{1}{10000}$
3. Cadastral maps	1 cm = 5 m to 500 m (1 : 5000 to 1 : 50000)	$\frac{1}{500}$ to $\frac{1}{50000}$
4. Location surveys	1 cm = 50 m to 200 m (1 : 5000 to 1 : 20000)	$\frac{1}{5000}$ to $\frac{1}{20000}$
5. Topographic surveys	1 cm = 250 m to 2500 m (1 : 25000 to 1 : 250000)	$\frac{1}{25000}$ to $\frac{1}{250000}$
6. Geographic maps	1 cm = 5000 m to 160000 m (1 : 500000 to 1 : 16000000)	$\frac{1}{500000}$ to $\frac{1}{16000000}$
7. Route Surveys	1 cm = 100 m (1 : 10000)	$\frac{1}{10000}$
8. Longitudinal sections		
(*i*) Horizontal scale	1 cm = 10 m to 200 m (1 : 1000 to 1 : 20000)	$\frac{1}{1000}$ to $\frac{1}{20000}$
(*ii*) Vertical scale	1 cm = 1 m to 2 m (1 : 100 to 1 : 200)	$\frac{1}{100}$ to $\frac{1}{200}$
9. Cross-sections (Both horizontal and vertical scales same)	1 cm = 1 m to 2 m (1 : 100 to 1 : 200)	$\frac{1}{100}$ to $\frac{1}{200}$

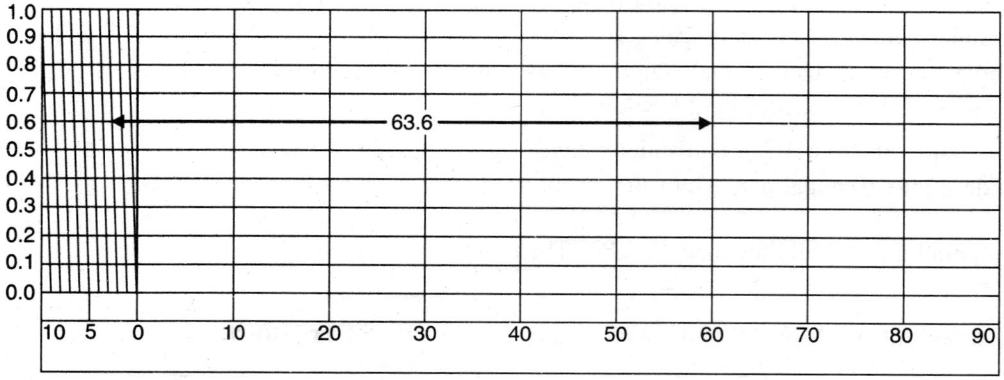

Fig. 2.8 Diagonal scale

IS 1562—1962 recommends diagonal scales *A*, *B*, *C*, and *D* as shown in Table 2.3.

Table 2.3 Indian standard diagonal scales (recommended)

Designation	*RF*	*Total graduated length*
A	$\frac{1}{1}$	150 cm
B	$\frac{1}{100000}$ $\frac{1}{50000}$ $\frac{1}{25000}$	100 cm
C	$\frac{1}{100000}$ $\frac{1}{50000}$ $\frac{1}{25000}$	50 cm
D	$\frac{1}{100000}$ $\frac{1}{8000}$ $\frac{1}{4000}$	150 cm

2.8 UNITS OF MEASUREMENTS

According to Standards of Weights and Measurements Act, India decided to give up FPS system used earlier and switched over to MKS in 1956. In 1960 system International (SI units) unit was approved by the conference of weights and measures. It is an international organisation of which most of the countries are the members. In this system also unit of linear measurement is metre. However, is this system use of centimeters and decameters are discouraged. Of course major difference between MKS and SI is in the use of unit of force. In MKS unit of force is kg-wt (which is commonly called as kg only)while in SI it is newton.

The recommended multipliers in SI units are given below

$$\text{Giga unit} = 1 \times 10^9 \text{ units}$$

$$\text{Mega unit} = 1 \times 10^6 \text{ units}$$

$$\text{Kilo unit} = 1 \times 10^3 \text{ units}$$

$$\text{unit} = 1 \times 10^0 \text{ units}$$

$$\text{Milli unit} = 1 \times 10^{-3} \text{ unit}$$

$$\text{Micro unit} = 1 \times 10^{-6} \text{ unit}$$

Commonly used linear units in surveying are kilometre, metre and millimetres. However centimetre is not yet fully given up.

For measuring angles sexagesimal system is used. In this system:

$$1 \text{ circumference} = 360°$$

$$1 \text{ degree} = 60' \text{ (minutes of arc)}$$

$$1 \text{ minute} = 60'' \text{ (seconds of arc)}$$

QUESTIONS

1. What is surveying? State its objects and uses.
2. Distinguish between geodetic surveying and plain surveying.
3. Explain the terms topographical surveying and cadastral surveying.
4. What are the fundamental principles of surveying? Explain briefly.
5. Discuss the classifications of surveying based on
   (*i*) instruments used
   (*ii*) objective of survey and
   (*iii*) methods employed.
6. Distinguish between plans and maps.
7. Explain with a neat sketch the construction of a diagonal scale to represent 1 cm = 5 m and show 53.6 m on it.

3

Linear Measurements in Surveying

All the distances required for making a plan are the horizontal distances. Hence in the field horizontal distances are measured or sufficient readings are taken to calculate horizontal distances. In this chapter the methods used for linear measurements are explained. Method of preparing a plan using only linear measurements is by conducting chain surveying. This method is also explained in this chapter and Indian Standard Conventions for showing objects on the map are presented at the end of the chapter.

3.1 METHODS OF LINEAR MEASUREMENTS

Various methods used for linear measurements may be grouped as

(*i*) Approximate

(*ii*) Using chain or tape

(*iii*) By optical means and

(*iv*) Using electromagnetic distance measurement instruments.

3.1.1 Approximates Methods of Linear Measurements

These methods are used in reconnaissance survey or to detect major mistakes committed while measuring with better methods. On smooth roads they can give results within 1 per cent error. These approximate measurements may be by:

(*i*) pacing

(*ii*) using passometer

(*iii*) using pedometer

(*iv*) using odometer or by

(*v*) using speedometer.

(*i*) **Pacing:** In this method surveyor walks along the line to be measured and counts the number of steps. Then the distance measured is equal to number of steps *X* average length of a step. Average length of a step can be found by walking along a known length. A normal man takes a step of length 0.75 m to 0.8 m.

(*ii*) **Using Passometer:** A passometer is a watch-like instrument which is carried vertically in the pocket of shirt or tied to a leg. It records number of steps taken. Thus the problem of counting number of steps is eliminated in this approximate method of linear measurement.

(*iii*) **Using Pedometer:** This instrument is similar to passometer but it can record the distance instead of number of steps. In this, zero setting and setting of step length is made before walking.

(*iv*) **Odometer:** This instrument is attached to the wheel of a cycle or other vehicle. It records the number of revolutions made by the wheel. Knowing the circumference of the wheel, the distance travelled may be found.

(*v*) **Speedometer:** Odometer calibrated to give distance directly is called speedometer. This is to be used for particular vehicle only. All automobiles are provided with speedometers. By running the vehicle along the line to be measured distance can be found.

3.1.2 Measurement with Chains or Tapes

Measurement of distances using chain or tape is termed as chaining. This is the accurate and commonly employed method in surveying: These instruments can be classified as (*i*) chain (*ii*) steel band and (*iii*) tapes.

(*i*) **Chains:** The chains are composed of 100 pieces of 4 mm diameter galvanised mild steel wires bent into rings at the end and joined to each other by three circular or oval shaped rings. These rings give flexibility to the chain. The ends of chains are provided with swivel joints (Ref. Fig. 3.1(*a*)),

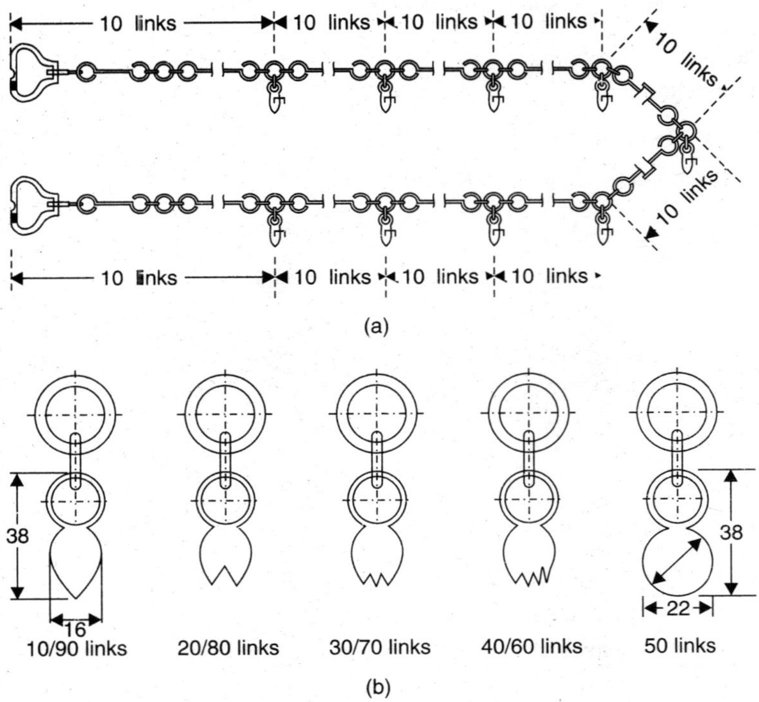

Fig. 3.1 Chain

so that the chain can be turned without twisting. To facilitate easy reading of the chain, brass tallies are provided. End of 10th link from each end is provided with a talley of one tooth, 20th link is provided with a talley of two teeth; 30th link with a talley of three teeth; 40th link with a talley of 4 teeth and the middle of chain is provided with a talley of circular shape [Ref. Fig. 3.1 (*b*)].

It is to be noted that

(*i*) length of a link is the distance between centres of two consecutive middle ring.

(*ii*) the length of the chain is from outside of one handle to the outside of the other handle.

Commonly used metric chains are of 20 m length. They have 100 links with talleys at every 2 m. Each link is of 0.2 m length. Simple rings are provided at every one metre length except wherever tallies are provided. The total length of chain is marked on the brass handle.

However 30 m chains are also in use. Length of each link is 0.3 m. It is not so convenient as 20 m chain to read, since no rings can be provided at one metre distance and each link needs multiplication with 0.3 to arrive at metre units. However as the influence of using 100 ft chain in olden days, this type of chain are also in market.

Steel Bond: It is also known as bond chain. It consists of steel of 12 to 16 mm width and 0.3 to 0.6 mm thickness. The steel ribbon is wound around an open steel cross or in a metal reel (Ref. Plate 3.1). Metric steel bands are available in lengths of 20 m and 30 m. Any one of the following two methods of markings are used:

(*i*) Providing brass studs at every 0.2 m and numbering at every metre. Last links from either end are subdivided in cm and mm.

(*ii*) Etching graduations as meters, decimeters and centimeters on one side of the band and 0.2 m links on the other side.

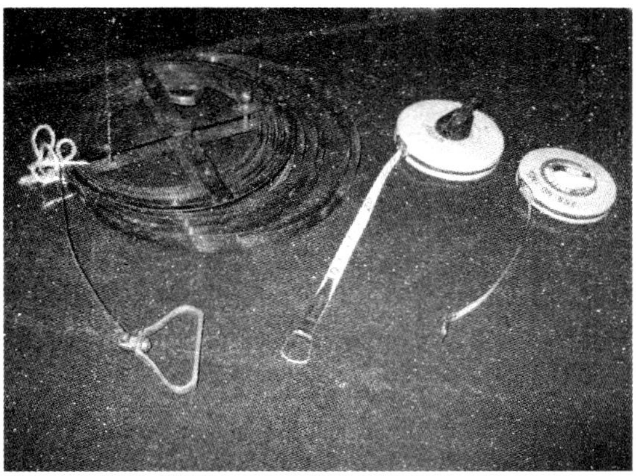

Plate 3.1

Tapes: Depending upon the materials used, they are classified as:

(*i*) cloth or linen tape

(*ii*) metallic tape

(*iii*) steel tape and

(*iv*) invar tape.

(*i*) **Cloth or Linen Tape:** 12 to 15 mm wide cloth or linen is varnished and graduations are marked. They are provided with brass handle at the ends. They are available in length of 10 m, 20 m, 25 m and 30 m. These tapes are light and flexible. However because of the following disadvantages they are not popular:

(*i*) Due to moisture they shrink.

(*ii*) Due to stretching they extend.

(*iii*) They are not strong.

(*iv*) They are likely to twist.

(*ii*) **Metallic Tape:** They are made up of varnished strip of waterproof linen interwooven with small wires of brass, copper or bronze. End 100 mm length of tapes are provided with leather or suitable strong plastic materials. Tapes of length 10 m, 20 m, 30 m and 50 m are available in a case of leather or corrosion resistant metal fitted with a winding device. Red and black coloured markings are used for indicating full metres and its fractions in centimetres. A typical metallic tape is shown is Fig. 3.2. These tapes are light, flexible and not easily broken. These tapes are commonly used in surveying.

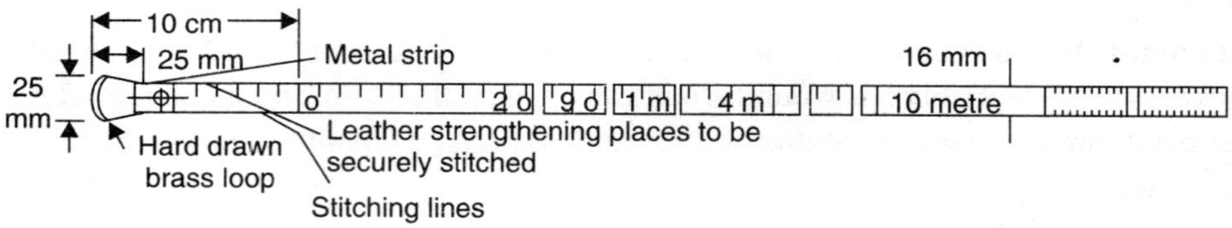

Fig. 3.2 Metallic tape

(*iii*) **Steel Tape:** A steel tape consists of 6 to 10 mm wide strip with metal ring at free end and wound in a leather or corrosion resistant metal case. It is provided with a suitable winding device. Tapes are marked indicating 5 mm, centimetres, decimetres and metres. The end 10 cm length is marked with millimetres also. 10 m, 20 m, 30 m, or 50 m tapes are used in surveying. Figure 3.3 shows a typical steel tape (Ref. Plate 3.1 also). Steel tapes are superior to metallic tapes as far as accuracy is concerned. However they are delicate. Care should be taken to wipe clean before winding. They should be oiled regularly to prevent corrosion.

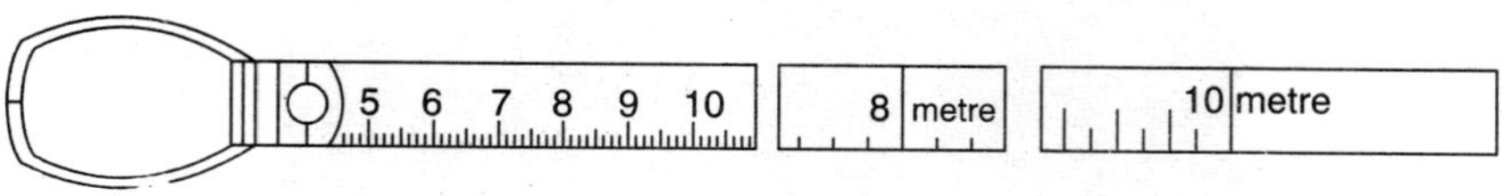

Fig. 3.3. Steel tape

(*iv*) **Invar Tape:** Invar is an alloy of nickel (36%) and steel. It's coefficient of thermal expansion is low. Hence errors due to variation in temperature do not affect measurements much. The width of tape is 6 mm. It is available in length 30 m, 50 m and 100 m. It is accurate but expensive.

3.1.3 Measurements by Optical Means

In this system, the telescope of the angle measuring instrument called theodolite (to be explained in Ch. 7) is provided with two additional cross hairs at *a* and *b* which are at distance '*i*' [Ref. Fig. 3.4]. To measure distance 'D' between two point P and Q instrument is set at P and a graduated staff is held vertically at Q and vertical intercept AB is recorded. Then distance D can be computed as explained as follows:

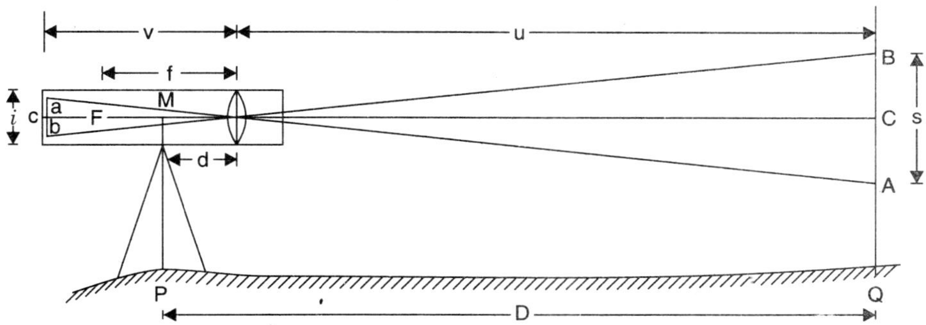

Fig. 3.4. Optical principle of linear measurement

$$\frac{v}{u} = \frac{i}{s} \quad \text{or} \quad u = v\,\frac{s}{i}$$

But from the law of optics,

$$\frac{1}{f} = \frac{1}{u} + \frac{1}{v}$$

Multiplying throughout by uf, we get

$$u = f + \frac{u}{v} f = f + \frac{s}{i} f = f + \frac{f}{i} s$$

If the distance between objective lence at O and centre of telescope is 'd' then

$$D = u + d$$

$$= f + \frac{f}{i} s + d$$

$$= ks + c \qquad \text{...(3.1)}$$

where $$k = \frac{f}{i} \text{ and } c = f + d$$

k and c are constants for a given instrument and hence can be found. Thus distance between P and Q can be found by measuring vertical intercept 's'. Thus distance is measured by optical means easity. This is called Tacheometric measurement. But this measurement is not that accurate as obtained by measuring with chain or tape. For details of this method of survey, reader has to refer specialised books on 'surveying and levelling'.

3.1.4 Linear Measurement with Electro-magnetic Distance Measurement Instruments

Now a days for measuring distances electronic equipments called electro-magnetic distance measurement (EDM) instruments have come in the market. They rely on the measurement of electromagnetic waves and measuring even fraction of wavelength by phase difference technique. They even display the distance measured. They are costly and delicate. Hence for ordinary surveying they are not used commonly. The details of such measurements are considered beyond the scope of this course.

3.2 INSTRUMENTS USED IN CHAINING

The following instruments are required for measurements with chain and tape:

(*i*) Arrows

(*ii*) Pegs

(*iii*) Ranging rods and ranging poles

(*iv*) Offset rods

(*v*) Laths

(*vi*) Whites

(*vii*) Plumb bobs and

(*viii*) Line ranger.

3.2.1 Arrows

When the length of the line to be measured is more than a chain length, there is need to mark the end of the chain length. Arrows are used for this purpose. A typical arrow is shown in Fig 3.5. Arrows are made up of 4 mm diametered steel wire with one end sharpened and other end bent into a loop. Length of an arrow is approximately 400 mm.

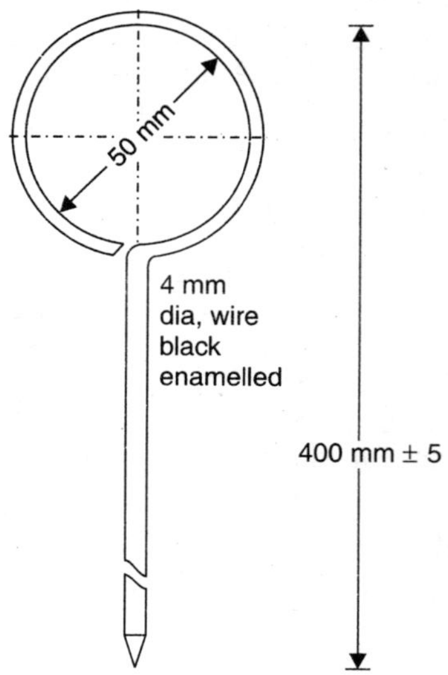

Fig. 3.5 Arrows

3.2.2 Pegs

Wooden pegs are used in measuring a length of a line to mark the end points of the line. The pegs are made of hard wood of 25 mm × 25 mm section, 150 mm long with one end tapered as shown in Fig. 3.6. When driven in ground to mark station points they project about 40 mm.

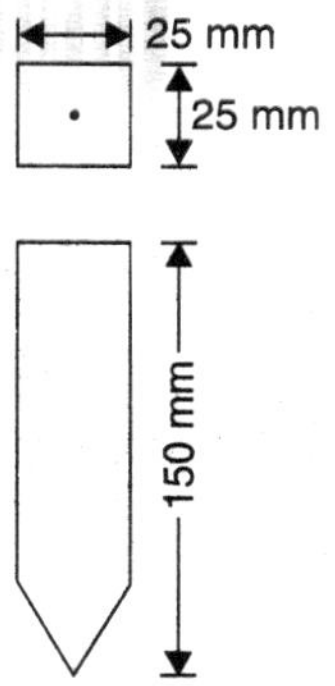

Fig. 3.6 Pegs

3.2.3 Ranging Rods and Ranging Poles

For ranging intermediate points along the line to be measured, ranging rods and ranging poles are used. Ranging rods are 2 to 3 m long and are made of hard wood. They are provided with iron shoe at one end as shown in Fig. 3.7.

They are usually circular is section with 30 mm diameter and are painted with 200 mm colour bands of red and white or with black and white. If distance is more than 200 m, for clear visibility they may be provided with multicoloured flags at their top. The ranging rods are occasionally used to measure short distances since they are painted with alternate colour of band 200 mm.

Ranging poles are similar to ranging rods except that they are longer. Their length varies from 4 m to 8 m and diameter from 60 mm to 100 mm. They are made of hard wood or steel. They are fixed in the ground by making 0.5 m holes and then packed to help them vertical.

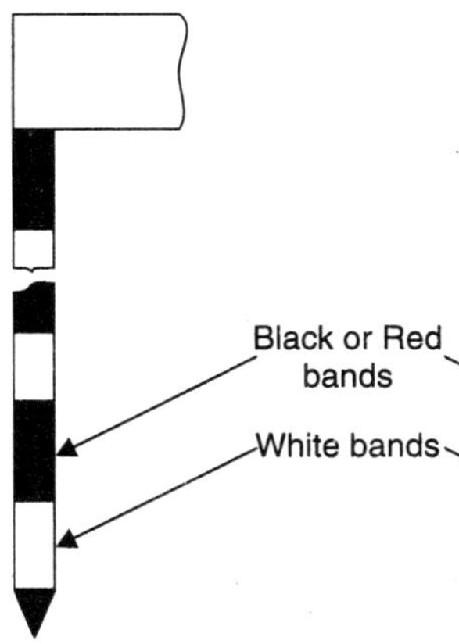

Fig. 3.7 Ranging rod

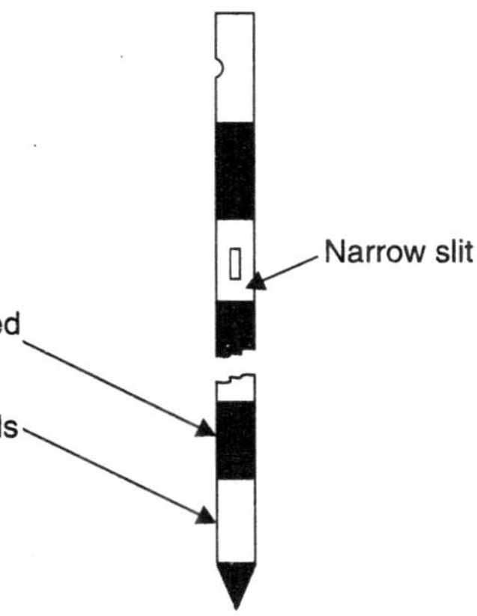

Fig. 3.8 Offset rod

3.2.4 Offset Rods

These rods are also similar to ranging rods and they are 3 m long. They are made up of hard wood and are provided with iron shoe at one end. A hook or a notch is provided at other end. At height of eye, two narrow slits at right angles to each other are also provided for using it for setting right angles. A typical offset rod is shown in Fig. 3.8.

3.2.5 Laths

Laths are 0.5 to 1.0 m long sticks of soft wood. They are sharpened at one end and are painted with white or light colours. They are used as intermediate points while ranging or while crossing depressions.

3.2.6 Whites

Whites are the pieces of sharpened thick sticks cut from the nearest place in the field. One end of the stick is sharpened and the other end is split. White papers are inserted in the split to improve the visibility. Whites are also used for the same purpose as laths.

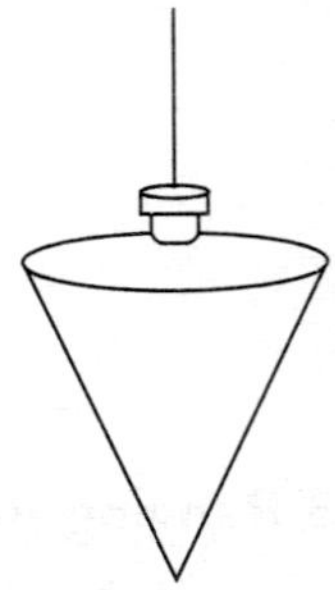

Fig. 3.9 Plumb bob

3.2.7 Plumb Bob

A typical plumb bob is shown in Fig. 3.9. In measuring horizontal distances along sloping ground plumb bobs are used to transfer the position to ground. They are also used to check the verticality of ranging poles.

3.2.8 Line Ranger

It is an optical instrument used for locating a point on a line and hence useful for ranging. It consists of two isosceless prisms placed one over the other and fixed in an instrument with handle. The diagonals of the prisms are silvered so as to reflect the rays.

To locate point C on line AB (ref. Fig. 3.10) the surveyor holds the instrument in hand and stands near the approximate position of C. If he is not exactly on line AB, the ranging rods at A and B appear separated as shown in Fig. 3.10 (*b*). The surveyor moves to and fro at right angles to the line AB till the images of ranging rods at A and B appear in a single line as shown in Fig. 3.10 (*c*). It happens only when the optical square is exactly on line AB. Thus the desired point C is located on the line AB.

Its advantage is it needs only one person to range. The instrument should be occasionally tested by marking three points in a line and standing on middle point observing the coincidence of the ranging rods. If the images of the two ranging rods do not appear in the same line, one of the prism is adjusted by operating the screw provided for it.

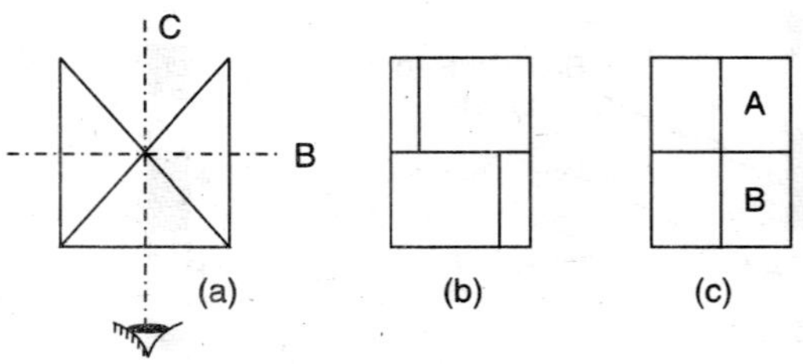

Fig. 3.10 Line ranger

3.3 CHAIN SURVEYING

Chain survey is suitable in the following cases:

(*i*) Area to be surveyed is comparatively small

(*ii*) Ground is fairly level

(*iii*) Area is open and

(*iv*) Details to be filled up are simple and less.

In chain surveying only linear measurements are made *i.e.* no angular measurements are made. Since triangle is the only figure that can be plotted with measurement of sides only, in chain surveying the area to be surveyed should be covered with a network of triangles. Figure 3.11 shows a typical scheme of covering an area with a network of triangles. No angle of the network triangles should be less than 30° to precisely get plotted position of a station with respect to already plotted positions of other station. As far as possible angles should be close to 60°. However, the arrangements of triangles to be adopted depends on the shape, topography, natural and artificial obstacles in the field.

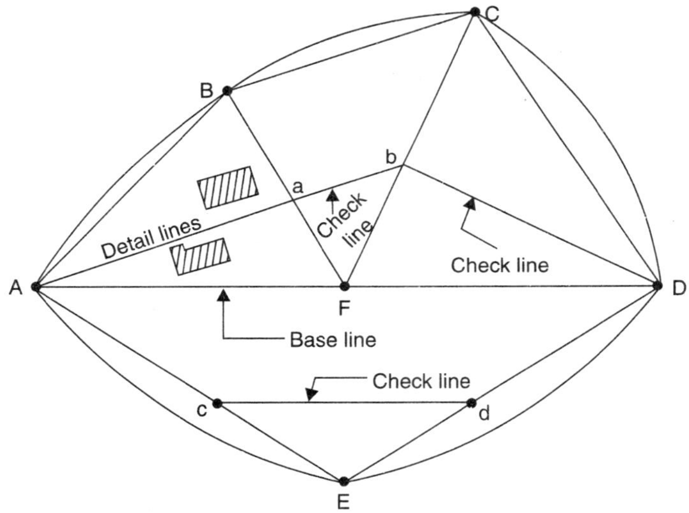

Fig. 3.11 Network of triangles

3.3.1 Technical Terms

Various technical terms used in connection with the network of the triangles in surveying are explained below:

Station: Station is a point of importance at the beginning or at the end of a survey line.

Main station: These are the stations at the beginning or at the end of lines forming main skeleton. They are denoted as A, B, C etc.

Subsidiary or tie stations: These are the stations selected on main lines to run auxiliary/secondary lines for the purpose of locating interior details. These stations are denoted as *a, b, c,* ...., etc., or as 1, 2, 3, ... etc.

Base line: It is the most important line and is the longest. Usually it is the line plotted first and then frame work of triangles are built on it.

Detail lines: If the important objects are far away from the main lines, the offsets are too long, resulting into inaccuracies and taking more time for the measurements. In such cases the secondary lines are run by selecting secondary stations on main lines. Such lines are called detail lines.

Check lines: These are the lines connecting main station and a substation on opposite side or the lines connecting to substations on the sides of main lines. The purpose of measuring such lines is to check the accuracy with which main stations are located.

3.3.2 Selection of Stations

The following points should be considered in selecting station points:

(*i*) It should be visible from at least two or more stations.

(*ii*) As far as possible main lines should run on level ground.

(*iii*) All triangles should be well conditioned (No angle less than 30º).

(*iv*) Main network should have as few lines as possible.

(*v*) Each main triangle should have at least one check line.

(*vi*) Obstacles to ranging and chaining should be avoided.

(*vii*) Sides of the larger triangles should pass as close to boundary lines as possible.

(*viii*) Tresspassing and frequent crossing of the roads should be avoided.

3.3.3 Offsets

Lateral measurements to chain lines for locating ground features are known as offsets. For this purpose perpendicular or oblique offsets may be taken (Ref. Fig. 3.12). If the object to be located (say road) is curved more number of offsets should be taken. For measuring offsets tapes are commonly used.

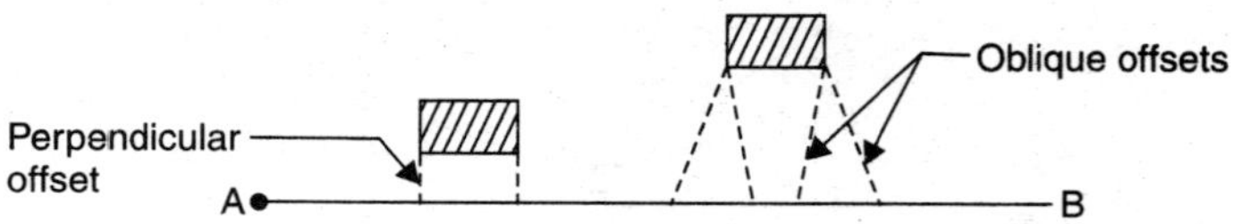

(a)

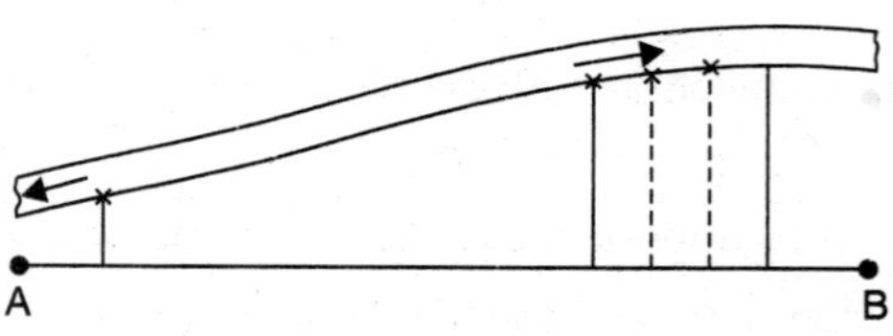

(b)

Fig. 3.12 Offsets

For setting perpendicular offsets any one of the following methods are used:

(*i*) Swinging

(*ii*) Using cross staffs

(*iii*) Using optical or prism square.

Perpendicular Offset by Swinging

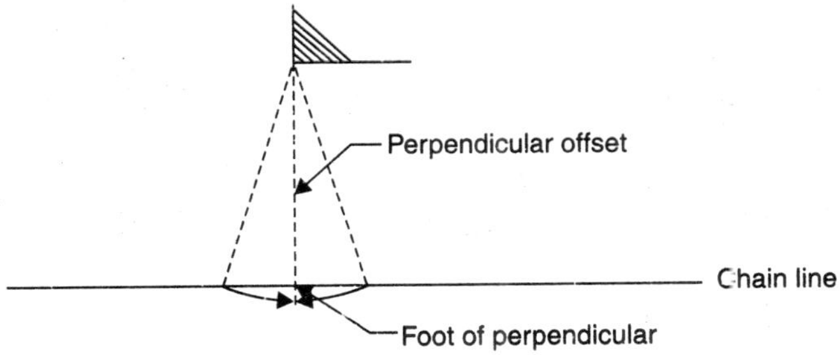

Fig. 3.13

Chain is stretched along the survey line. An assistant holds the end of tape on the object. Surveyor swings the tape on chain line and selects the point on chain where offset distance is the least (Fig. 3.13) and notes chain reading as well as offset reading in a field book on a neat sketch of the object.

Perpendicular Offsets Using Cross Staffs

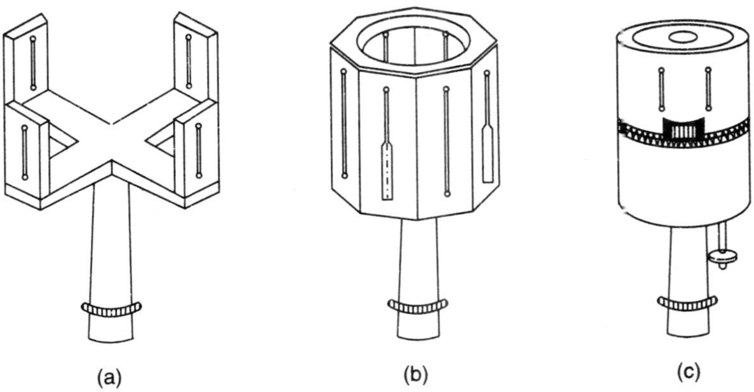

Fig. 3.14 Cross staff

Figure 3.14 shows three different types of cross staffs used for setting perpendicular offsets. All cross staffs are having two perpendicular lines of sights. The cross staffs are mounted on stand. First line of sight is set along the chain line and without disturbing setting right angle line of sight is checked to locate the object. With open cross staff (Fig. 3.14 (*a*)) it is possible to set perpendicular only, while with french cross staff (Fig. 3.14 (*b*)), even 45° angle can be set. Adjustable cross staff can be used to set any angle also, since there are graduations and upper drum can be rotated over lower drum.

Perpendicular Offsets Using Optical Square and Prism Square

These instruments are based on the optical principle that if two mirrors at angle 'θ' to each other reflect a ray at angle '2θ'. Figure 3.15 shows a typical optical square.

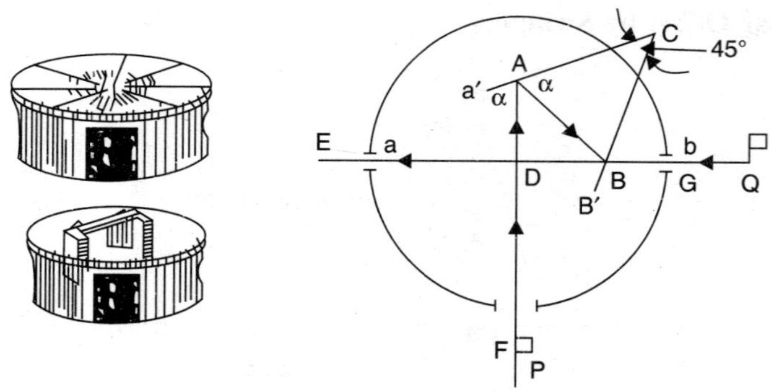

Fig. 3.15 Optical square

Optical square consists of a metal box about 50 mm in diameter and 125 mm deep. In the rim of the box there are three openings:

(*i*) a pin hole at E

(*ii*) a small rectangular slot at G, and

(*iii*) a large rectangular slot at F.

A and B are the two mirrors placed at 45° to each other. Hence the image of an object at F which falls on A gets reflected and emerge at E which is at right angles to the line FA. The mirror A which is opposite to the opening at F is fully silvered. It is fitted to a frame which is attached to the bottom plate. If necessary this mirror can be adjusted by inserting a key on the top of the cover. The mirror B which is in the line with EG is silvered in the top half and plain in the bottom half. It is firmly attached to the bottom plate of the box.

The ranging rod at Q is directly sighted by eye at E in the bottom half of the B which is a plain glass. At the same time in the top half of B, the reflected ray of the object at P is sighted. When the image of P is in the same vertical line as the object at Q, then the lines PA is at right angles to the line EB.

This instrument can be used for finding foot of the perpendicular or to set a right angle.

In prism square, instead of two mirrors at 45° to each other a prism which has two faces at 45° to each other is used [Fig. 3.16.]. Its advantage is it will not go out of adjustment even after long usage.

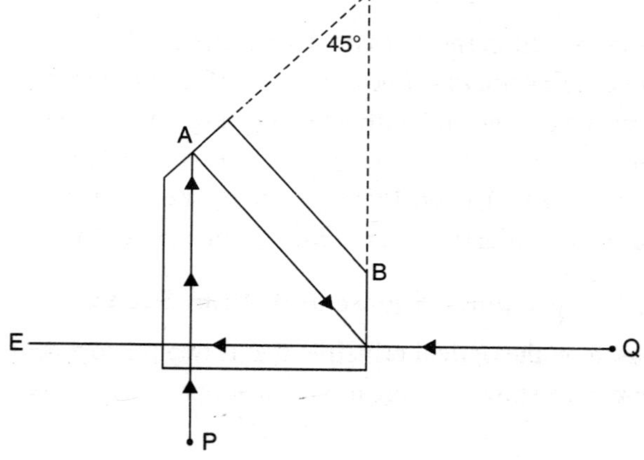

Fig. 3.16 Prism square

3.3.4 Field Book

All observations and measurements taken during chain surveying are to be recorded in a standard field book. It is a oblong book of size 200 mm × 120 mm, which can be carried in the pocket.

There are two forms of the book (*i*) single line and (*ii*) double line. The pages of a single book are having a red line along the length of the paper in the middle of the width. It indicates the chain line. All chainages are written across it. The space on either side of the line is used for sketching the object and for noting offset distances. In double line book there are two blue lines with a space of 15 to 20 mm is the middle of each book. The space between the two lines is utilised for noting the chainages. Figure 3.17 shows typical pages of a field books.

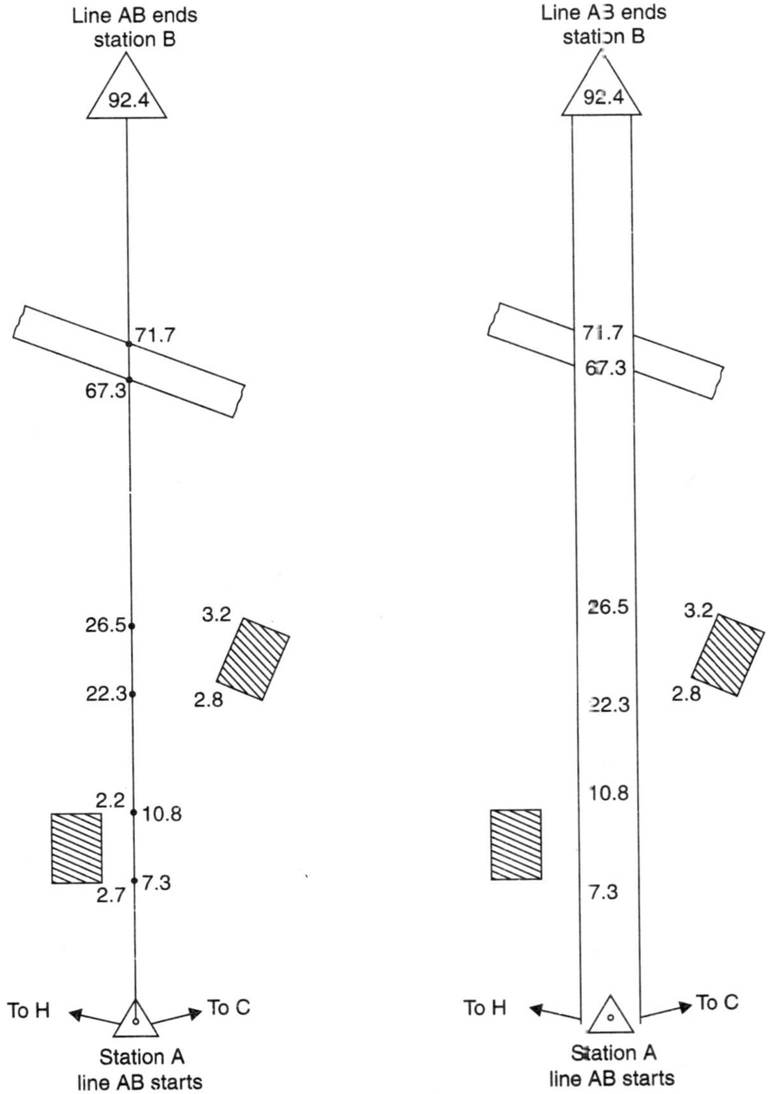

Fig. 3.17

3.3.5 Field Work

As soon as the survey party arrives in the field the following details are entered in the field book:

(*i*) Title of the survey work

(*ii*) The date of survey

(*iii*) The names of the members of the party.

The field work may be divided into the following:

(*i*) Reconnaissance survey.

(*ii*) Marking stations, drawing reference sketches.

(*iii*) Line by line surveying.

Reconnaissance survey consists in going round the field and identifying suitable stations for the network of triangles. Neat sketch of network is drawn and designated. The typical key plan drawn is similar to one shown in Fig. 3.11.

All main stations should be marked on the ground. Some of the methods used for marking are:

(*a*) Fixing ranging poles

(*b*) Driving pegs

(*c*) Marking a cross if ground is hard

(*d*) Digging and fixing a stone.

Then reference sketches are drawn in the field book so as to identify stations when the development works are taken up. For this measurements with respect to three permanent points are noted. The permanent points may be

(*a*) Corner of a building

(*b*) Posts of gates

(*c*) Corners of compound walls

(*d*) Electric poles

(*e*) A tree.

After that, line by line surveying is conducted to locate various objects with respect to chain lines.

3.3.6 Office Work

It consists in preparing the plan of the area to a suitable scale making use of measurements and sketches noted in the field book.

3.4 RANGING

When a survey line is longer than a chain length, it is necessary to align intermediate points on chain line so that the measurements are along the line. The process of locating intermediate points on survey line is known as ranging. There are two methods of ranging viz., direct ranging and reciprocal ranging.

3.4.1 Direct Ranging

If the first and last points are intervisible this method is possible. Figure 3.18 shows the intervisible stations AB in which an intermediate point C is to be located. Point C is selected at a distance slightly less than a chain length. At points A and B ranging rods are fixed. The assistant holds another ranging rod near C. Surveyor positions himself approximately 2 m behind station A and looking along line AB directs the assistant to move at right angles to the line AB till he aligns the ranging rod along AB. Then surveyor instructs the assistant to mark that point and stretch the chain along AC.

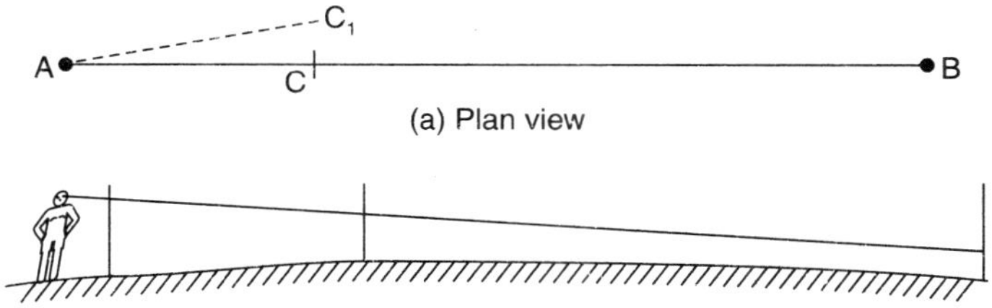

(a) Plan view

(b) Sectional view

Fig. 3.18 Direct ranging

3.4.2 Indirect or Reciprocal Levelling

Due to intervening ground, if the ranging rod at B is not visible from station A, reciprocal ranging may be resorted. Figure 3.19 shows this scheme of ranging. It needs two assistants one at point M and another at point N, where from those points both station A and station B are visible. It needs one surveyor at A and another at B. To start with M and N are approximately selected, say M_1 and N_1. Then surveyor near end A ranges person near M to position M_2 such that AM_2N_1 are in a line. Then surveyor at B directs person at N, to move to N_2 such that BN_2M_2 are in a line. The process is repeated till AMNB are in a line.

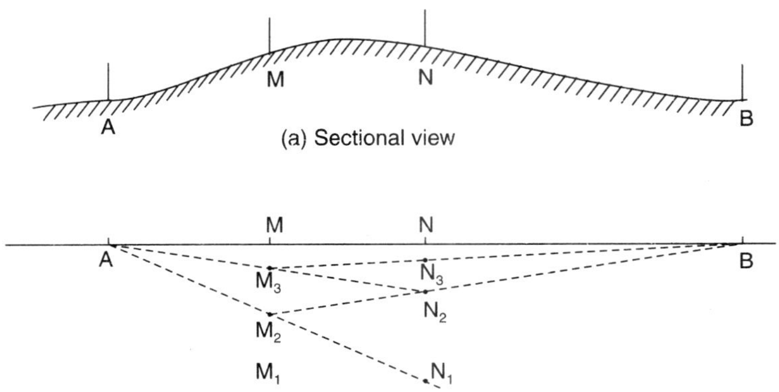

(a) Sectional view

(b) Plan view

Fig. 3.19 Reciprocal ranging

3.5 OBSTACLES IN CHAINING

Though it is desirable to select stations so as to avoid obstacles, occasionally the obstacles are unavoidable.

Various obstacles to chaining may be grouped into:

(*i*) Obstacles to ranging (chaining free-vision obstructed)
(*ii*) Obstacles to chaining (chaining obstructed-vision free)
(*iii*) Obstacles to both ranging and chaining.

Various methods of overcoming these obstacles are explained is this article.

3.5.1 Obstacles to Ranging

These obstacles can be further classified into the following categories:

(*a*) Both ends of the line are visible from some intermediate points. Intervening ground is an example of such obstacle. By resorting to reciprocal ranging this difficulty can be overcome.

(*b*) Both ends of the line may not be visible from intermediate points on the line, but may be visible from a point slightly away from the line. Intervening trees and bushes are the examples of such obstacles. This obstacle to chaining may be overcome by measuring along a random line as shown in Fig. 3.20. In this case required length

$$AB = \sqrt{EC^2 + CB^2}$$

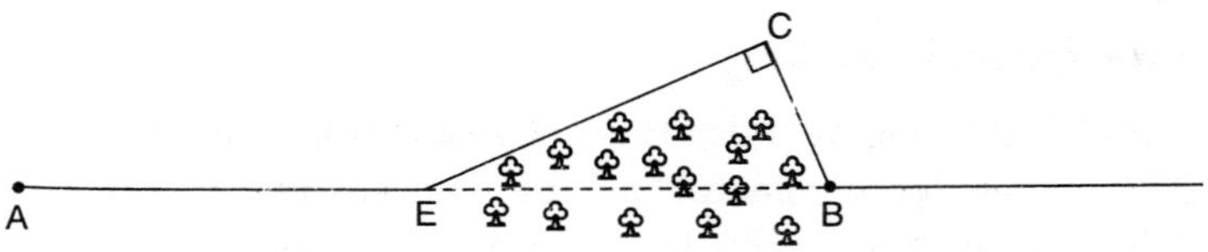

Fig. 3.20 Obstacle to ranging

3.5.2 Obstacles to Chaining

In this type the ends of lines are visible but chaining is obstructed. Examples of such obstructions are ponds, lakes, marshy land etc. Various geometric properties may be used to find obstructed length CB as shown in Fig. 3.21.

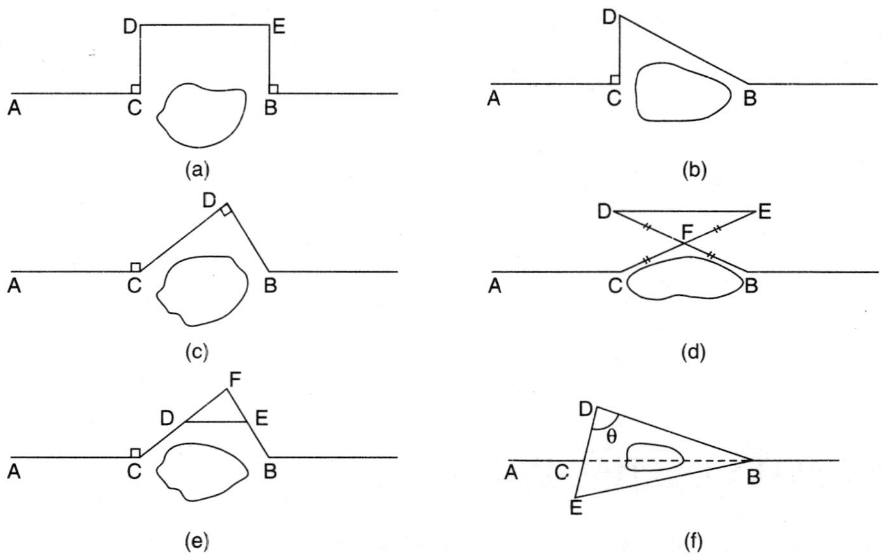

Fig. 3.21 Obstacles to chaining

(*a*) Set CD and BE perpendiculars to AB, such that CD = BE. Then

$$CB = DE \qquad \text{[Fig 3.22 (a)]}$$

(*b*) Set perpendicular CD to AB. Measure CD and DB. Then

$$CB = \sqrt{BD^2 - CD^2} \qquad \text{[Fig. 3.22 (b)]}$$

(*c*) Set CD and DB such that DB ⊥ CD. Measure them. Then

$$CB = \sqrt{CD^2 + BD^2} \qquad \text{[Fig. 3.22 (c)]}$$

(*d*) Select a convenient point F. Set FE = CF and FD = BF. Then

$$CB = DE \qquad \text{[Fig. 3.22 (d)]}$$

(*e*) Select a convenient point F. Locate D and E such that CF = *n* DF and BF = *n* EF. Measure DE. Then,

$$\frac{CF}{DF} = \frac{BF}{EF} = n = \frac{CB}{DE}$$

$$\therefore \quad CB = n\ DE \qquad \text{[Fig. 3.22 (e)]}$$

(*f*) Select points D and E on line passing through C [Fig. 3.22 (*b*)]. Measure CD, CE, DB and EB. Then, from ΔBDE,

$$\cos\theta = \frac{BD^2 + DE^2 - EB^2}{2BD \cdot DE} \qquad ...(a)$$

and from ΔBDC,

$$\cos\theta = \frac{CD^2 + BD^2 - CB^2}{2CD \cdot BD} \qquad ...(b)$$

From eqn. (*a*), cos θ can be found and substituting it in eqn. (*b*), the obstructed length CB can be found.

3.5.3 Obstacles to Both Chaining and Ranging

Building is a typical example of this obstacle. Referring to Fig. 3.22, line AB is to be continued beyond the obstacle, say as GH. Four possible methods are presented below:

(*a*) Set perpendiculars AC, BD such that AC = BD [Fig. 3.22 (*a*)]. Extend line CD to F. Drop perpendiculars EG and FH to line CF such that EG = FH = AC. GH is the continuation of line AB and DE = BG.

(*b*) Referring to Fig. 3.22 (*b*), set BC ⊥ to AB. Select D on extended line of AC. Set perpendicular DH such that AD = DH. Select point E on DH such that DE = DC. Then arcs of length EG = BC and arc of length HG = AB are drawn from E and H respectively and G is located. GH is continuation of AB and BG = CE.

(*c*) Referring to Fig. 3.22 (*c*), C is located such that AC = BC = AB. Extend AC to D and construct equilateral triangle DEF. Extend DF to H such that DH = DA. Locate convenient point I on

HD and construct equilateral triangle to locate G. Then GH is the continuation of line AB and length BG is given by

$$BG = AH - AB - GH = AD - AB - GH$$

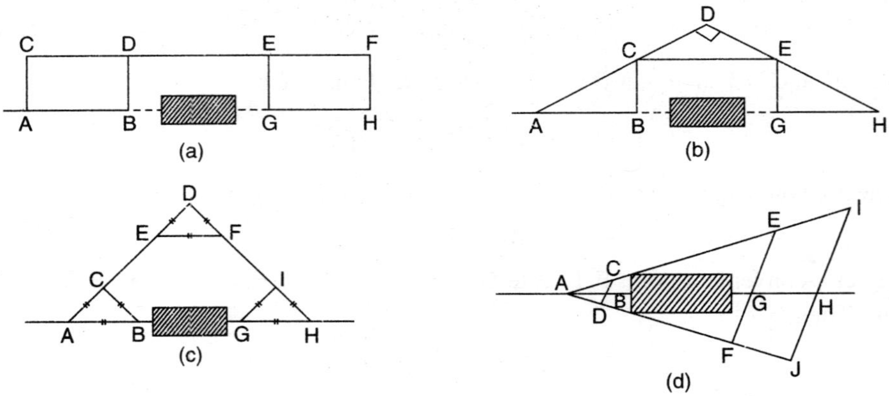

Fig. 3.22 Obstacles to both ranging and chaining

(*d*) In the method shown in Fig. 3.22 (*d*), points C and D are selected such that CBD is in a line. Extend AC to E and I such that $AE = n \times AC$ and $AI = m \times AC$. Similarly Extend AD to F and J such that, $AF = n \times AD$ and $AJ = m \times AD$. Locate G and H on lines EF and IJ such that, $EG = n \times BC$ and $IH = m \times BC$. Then GH is the continuation of line AB. Now,

$$AG = n \times AB$$

$$\therefore \quad BG = n \times AB - AB = (n - 1)\, AB$$

■ **Example 3.1:** *In chaining past a pond, stations C and B were taken on the opposite sides of the pond. A line DCE was set by selecting CD = 220 m and CE = 280 m. The lines DB and ED which are on the opposite sides of the pond are measured. If DB = 500 m and EB = 600 m, find obstructed length CD.*

Solution: Referring to Fig. 3.21 (*f*),

$$CD = 220 \text{ m} \qquad CE = 280 \text{ m}$$

$$DB = 500 \text{ m} \qquad EB = 600 \text{ m}$$

$$\therefore \quad DE = DC + CE = 220 + 280 = 500 \text{ m}$$

From ΔBDE,

$$\cos\theta = \frac{DE^2 + BD^2 - EB^2}{2\, DE \cdot BD} = \frac{500^2 + 500^2 - 600^2}{2 \times 500 \times 500} = 0.28$$

From ΔBDC,

$$\cos\theta = \frac{CD^2 + BD^2 - BC^2}{2\, CD \cdot BD}$$

$$0.28 = \frac{220^2 + 500^2 - BC^2}{2 \times 220 \times 500}$$

$$61600 = 220^2 + 500^2 - BC^2$$

$\therefore$ **BC = 486.62 m Ans.**

3.6 ERRORS IN CHAINING

Errors in chaining may be classified as:

(*i*) Personal errors

(*ii*) Compensating errors, and

(*iii*) Cumulating errors.

3.6.1 Personal Errors

Wrong reading, wrong recording, reading from wrong end of chain etc., are personal errors. These errors are serious errors and cannot be detected easily. Care should be taken to avoid such errors.

3.6.2 Compensating Errors

These errors may be sometimes positive and sometimes negative. Hence they are likely to get compensated when large number of readings are taken. The magnitude of such errors can be estimated by theory of **probability**. The following are the examples of such errors:

(*i*) Incorrect marking of the end of a chain.

(*ii*) Fractional part of chain may not be correct though total length is corrected.

(*iii*) Graduations in tape may not be exactly same throughout.

(*iv*) In the method of stepping while measuring sloping ground, plumbing may be crude.

3.6.3 Cumulative Errors

The errors, that occur always in the same direction are called cumulative errors. In each reading the error may be small, but when large number of measurements are made they may be considerable, since the error is always on one side. Examples of such errors are:

(*i*) Bad ranging

(*ii*) Bad straightening

(*iii*) Erroneous length of chain

(*iv*) Temperature variation

(*v*) Variation in applied pull

(*vi*) Non-horizontality

(*vii*) Sag in the chain, if suspended for measuring horizontal distance on a sloping ground.

Errors (*i*), (*ii*), (*vi*) and (*vii*) are always +ve since they make measured length more than actual. Errors (*iii*), (*iv*) and (*v*) may be +ve or –ve.

3.7 TAPE CORRECTIONS

The following five corrections may be found for the measured lengths of tape:

(*i*) Corrections for absolute length

(*ii*) Corrections for pull

(*iii*) Corrections for temperature

(*iv*) Corrections for slope and

(*v*) Corrections for sag.

3.7.1 Corrections for Absolute Length

Let, l = designated length of tape, l_a = actual length of tape.

Then correction for chain length

$$c = l_a - l$$

Hence, if the total length measured is L, the total correction

$$C_a = \frac{c}{l} L \qquad ...(3.1)$$

∴ Corrected length = L + C_a

If A is the measured area with incorrect tape, the correct area

$$= \left(1 + \frac{c}{l}\right)^2 A = \left(1 + \frac{2c}{l}\right) A, \text{ since C is small.} \qquad ...(3.2)$$

3.7.2 Corrections for Pull

If pull applied while standardising the length of tape and pull applied in the field are different, this correction is required.

Let, P_0 = Standard pull

P = Pull applied in the field

A = Cross-sectional area of the tape

L = Measured length of line.

E = Young's modulus of the material of tape, then

$$C_p = \frac{(P - P_0) L}{AE} \qquad ...(3.3)$$

The above expression takes care of sign of the correction also.

3.7.3 Correction for Temperature

Let T_0 = Temperature at which tape is standardised

T_m = Mean temperature during measurement

α = Coefficient of thermal expansion of the material of the tape and

L = Measured length,

Then the temperature correction C_t is given by,

$$C_t = L\alpha\,(T_m - T_0) \qquad ...(3.4)$$

The above expression takes care of sign of the correction also.

3.7.4 Correction for Slope

If the length measured is 'L' and the difference in the levels of first and the last point is 'h', then slope correction

$$C_{sl} = L - \sqrt{L^2 - h^2}$$

$$= L\,[1 - \sqrt{1-(h/2)^2}\,]$$

$$= L\left[1-\left(1-\frac{h^2}{2L^2}-\frac{h^4}{8L^4}....\right)\right]$$

$$= \frac{h^2}{2L} \qquad ...(3.5)$$

If measured length is L and slope is 'θ', then

$$C_{sl} = L - L\cos\theta = L = (1-\cos\theta) \qquad ...(3.6)$$

This correction is always –ve.

3.7.5 Correction for Sag

While measuring on unevenly sloping ground, tapes are suspended at shorter length and horizontal distances are measured. This technique eliminates errors due to measurement along slopes, but necessitates correction for sag [Fig. 3.23]. Hence, measured length is more than actual length. Thus the correction is –ve. The correction, which is difference between the length of catenary and true length is given by

$$C_s = \frac{1}{24}\left(\frac{W}{P}\right)^2 L \qquad ...(3.7)$$

where, W = the weight of the tape of span length

P = the pull applied

and L = measured length

It may be noted that if pull is more than standard pull, the correction for pull is +ve, while correction for sag is always –ve. The pull for which these two corrections neutralise each other is called 'normal tension'. Hence normal tension R_n may be found as,

$$C_p = C_s$$

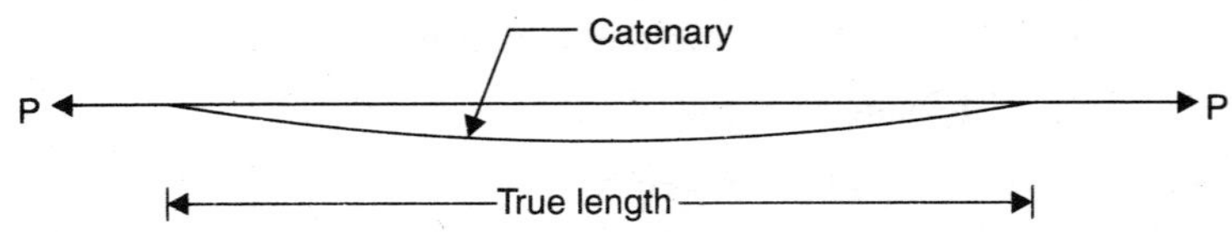

Fig. 3.23

$$\frac{(P_n - P_0)L}{AE} = \frac{1}{24}\left(\frac{W}{P_n}\right)^2 L$$

$$(P_n - P_0)\,P_n^2 = \frac{1}{24}\,W^2\,AE$$

or
$$P_n = \frac{0.204\,W\sqrt{AE}}{\sqrt{P_n - P_0}} \qquad ...(3.8)$$

The value of P_n is to be determined by trial and error method.

■ **Example 3.2:** *A distance of 1500 m was measured with a 20 m chain. After the measurement chain was found to be 80 mm longer. If the length of chain was perfectly correct while starting measurement, what is the true length of the line measured?*

Solution: Average correction per chain length

$$c = \frac{0 + 80}{2} = 40 \text{ mm} = 0.04 \text{ m}$$

Chain length = 20 m

∴ Correction for measured length

$$C_a = L\,\frac{c}{l} = 1500\,\frac{0.04}{20}$$

$$= 3.0 \text{ m}$$

∴ **True length** = 1500 + 3.0 = **1503 m** **Ans.**

■ **Example 3.3:** *A survey was conducted with a 20 m chain and plan of the field was drawn to a scale of 1 cm = 5 m. The area of the plan was found to be 62.8 cm². However when the chain was tested at the end of work, it was wound to be 20.10 m. Assuming the length was exactly 20.0 m in the beginning of survey work, determine the true area of the field.*

Solution: Initial length of the chain = 20 m

Length at the end of work = 20.1 m

∴ Average length of chain $= \frac{20 + 20.1}{2} = 20.05$ m

∴ Average correction per chain length = 20.05 – 20.00 = 0.05 m

Measured area on plan = 62.8 cm^2

Scale 1 cm = 5 m

∴ Measured area on field = 62.8×5^2

$= 1570 \text{ m}^2$

∴ **Corrected area on the ground** = $A\left(\frac{l^1}{l}\right)^2 = 1570\left(\frac{20.05}{20.00}\right)^2 = \mathbf{1577.860\ m^2}$ **Ans.**

■ **Example 3.4:** *A 20 m tape was used for measuring a line at an average temperature of 65° F. The measured distances on the ground and slope of the ground are as given below:*

2°	*18′*	*for*	*125 m*
3°	*30′*	*for*	*250 m*
1°	*42′*	*for*	*a distance of 170 m*

If the temperature at which tape was standardised is 80° F, find the true length of the line. Take $\alpha = 6.2 \times 10^{-6}/°F$.

Solution: Measured horizontal distance = $\Sigma L \cos\theta$

$= 125 \cos 2°\ 18' + 250 \cos 3°\ 30' + 170 \cos 1°\ 42'$

$= 544.367$ m

Temperature correction

$$C_t = L\alpha\,(T_m - T_0)$$

$$= 544.367 \times 6.2 \times 10^{-6}\,(65 - 80)$$

$$= -0.051 \text{ m}$$

Correct horizontal length

$= 544.367 - 0.051 =$ **544.316 m Ans.**

■ **Example 3.5:** *To measure a base line, a steel tape 30 m long, standardised at 15° C with a pull of 80 N was used. Find the correction per tape length, if the temperature at the time of measurement is 25° C and the pull exerted is 150 N.*

Take Young's modulus $E = 2 \times 10^5$ *N/mm²* *and coefficient of thermal expansion* $\alpha = 11.2 \times 10^{-6}/°C$. *Cross-sectional area of tape is 8 mm².*

Solution: $l = 30$ m, $\alpha = 11.2 \times 10^{-6}/°C$, $T_0 = 15°C$, $P_0 = 80$ N

$T_m = 25°C$, $P = 150$ N

∴ Correction for temperature

$$C_t = l\,\alpha\,(T_m - T_0)$$

$$= 30 \times 11.2 \times 10^{-6}\,(25 - 15)$$

$$= 3.360 \times 10^{-3} \text{ m}$$

Correction for pull

$$C_p = \frac{(P - P_0)\,l}{AE}$$

$$= \frac{(150 - 80) \times 30}{2 \times 10^5 \times 8}$$

$$= 1.3125 \times 10^{-3} \text{ m}$$

[*Note:* Unit of AE comes out to be Newton only, if A is in cm^2 or mm^2 and E in N/cm^2 or N/mm^2].

∴ Total correction for temperature and pull

$$= C_t + C_p = 3.360 \times 10^{-3} + 1.3125 \times 10^{-3}$$

$$= 4.6725 \times 10^{-3} \text{ m per chain length}$$

■ **Example 3.6:** *Calculate sag correction for a 30 m steel tape under a pull of 80 N, if it is suspended in three equal spans. Unit weight of steel is 78.6 kN/m^3. Area of cross-section of tape is 8 mm^2.*

Solution: Length of each span = 10 m

$$W = \text{wt. of taper per span length}$$

$$= 78.6 \times 10 \times (8 \times 10^{-6})$$

$$= 6288 \times 10^{-6} \text{ kN} = 6.288 \text{ N}$$

[*Note:* 1 mm^2 = (0.001)^2 m^2 = 1 × 10^{-6} m^2

P = 80 N L = 10 m

∴ Correction for each span

$$= \frac{1}{24}\left(\frac{6.288}{80}\right)^2 \times 10$$

$$= 2.574 \times 10^{-3} \text{ m}$$

∴ **Correction for three spans**

$$= 3 \times 2.574 \times 10^{-3} \text{ m}$$

$$= \mathbf{7.722 \times 10^{-3} \text{ m}} \quad \textbf{Ans.}$$

■ **Example 3.7:** *A 30 m steel tape was standardised under 60 N pull at 65° F. It was suspended in 5 equal span during measurement. The mean temperature during measurement was 90° F and the pull exerted was 100 N. The area of the cross-section of the tape was 8 mm^2. Find the true length of the tape, if,*

$\alpha = 6.3 \times 10^{-6}$/°F, $E = 2 \times 10^5$ N/mm^2 and unit weight of steel = 78.6 kN/m^3.

Solution: *Correction for temperature:*

$$C_t = l\,\alpha\,(T_m - T_0)$$

$$= 30 \times 6.3 \times 10^{-6}\,(90 - 65)$$

$$= 4.725 \times 10^{-3} \text{ m}$$

Correction for pull:

$$C_p = \frac{(P - P_0)\,l}{AE} = \frac{(100 - 60) \times 30}{8 \times 2 \times 10^5} = 0.75 \times 10^{-3} \text{ m}$$

Correction for sag:

Each span length $= \frac{30}{5} = 6$ m

Weight of tape per span, W = $78.6 \times 6 \times 8 \times 10^{-6}$

$= 3772.8 \times 10^{-6}$ kN

$= 3.7728$ N

∴ Correction for sag per span

$$= \frac{1}{24}\left(\frac{3.7728}{100}\right)^2 \times 6$$

$= 0.3559 \times 10^{-3}$ m

∴ Corrections for 5 spans

$C_s = 5 \times 0.3559 \times 10^{-3}$

$= 1.779 \times 10^{-3}$ m (– ve)

Noting that correction C_t and C_p are +ve

While correction C_s is –ve, we get

Total correction $= 4.725 \times 10^{-3} + 0.75 \times 10^{-3} - 1.779 \times 10^{-3}$

$= 3.696 \times 10^{-3}$ m

$= 3.696$ mm

∴ **True length of tape**

$= 30 + 3.696 \times 10^{-3}$

= 30.003696 m **Ans.**

3.8 CONVENSIONAL SYMBOLS

IS 962—1989, 'code of practice for architectural and building drawings' has specified standard symbols for various objects as shown in Table 3.1 on next page.

If coloured plans are to be made, the code recommends light washes of the following shades:

For roads – Burnt sienna

For buildings – Light grey

For compound walls – Indigo

For water – Borders edged with Prussian blue

For trees – Green.

Table 3.1

Feature	Symbol	Feature	Symbol
Chain line		Road under railway	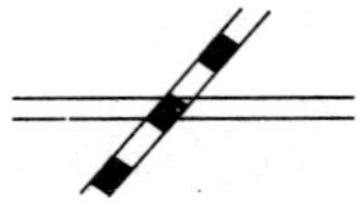
Triangulation station		Boundaries without pillars	
		Boundaries with pillars	
Traverse station	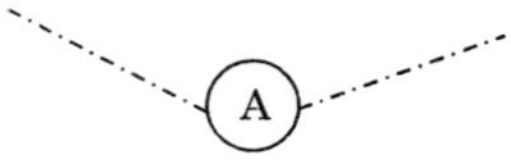	Township or taluka boundaries	
Building		River	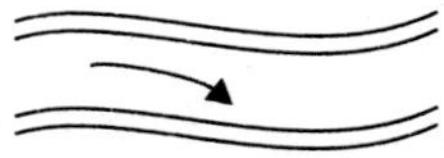
Shed with open side			
Shed with closed side		Pond	
Temple, mosque and church		Electric line	
Path			
Unfenced road		Tree	
Fenced road		Orchard	
Railway line: Single	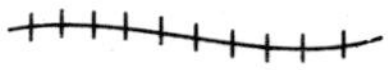	Woods	
Railway line: Double	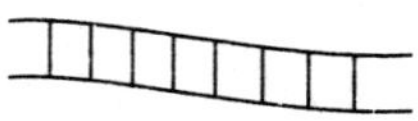	Grass	
Road bridge	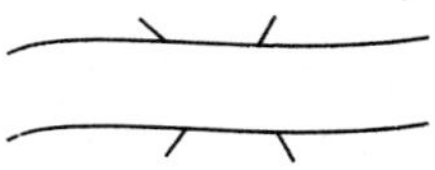	Cutting	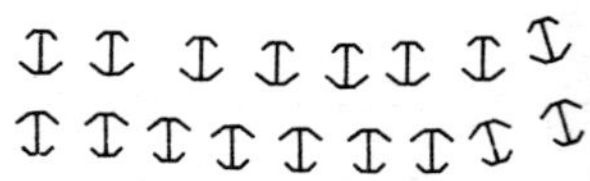
Level crossing	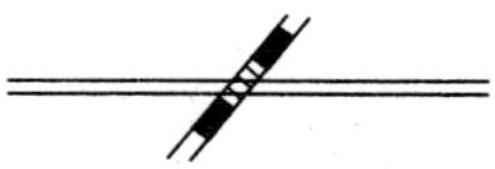	Embankment	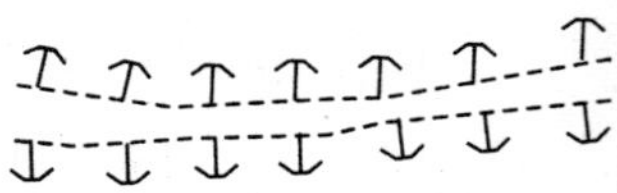
Road over railway	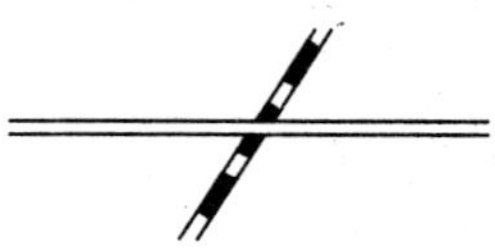	North line	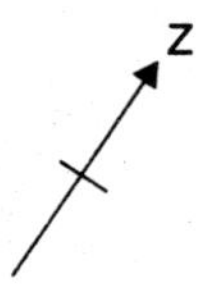

QUESTIONS

1. Briefly explain the various approximate methods of measurement of distances.
2. Briefly explain features of metric chain.
3. Distinguish between metallic tape and steel tape.
4. Write short notes on
   (*i*) Steel band (*ii*) Invar tape
   (*iii*) Ranging rods (*iv*) Line ranger.
5. Explain the terms
   (*i*) Check line and tie line (*ii*) Reference sketch.
6. List the various points to be considered in selecting stations for surveying.
7. Explain the use and working of
   (*i*) Open cross staff (*ii*) Prism square.
8. Explain the method of reciprocal levelling. When do you need it?
9. Explain any two methods of overcoming chaining problem in the following cases:
   (*i*) the obstruction is in the form of a pond.
   (*ii*) the obstruction is in the form of a building.
10. There is an obstacle in the form of a pond on the main chain line AB. Two points C and D were taken on the opposite sides of the pond. On the left of CD, a line CE was laid out 65 m in length and a second line CF of 85 long was laid on the right of CD such that ECF arc in a line. Determine the obstructed length CD. Given ED = 110 m and DF = 120 m. [**Ans.** CD = 87.015 m]
11. Distinguish between comulative error and compensating errors. Give examples for each case.
12. A 30 m tape used for measuring a line was found to be 30.01m at the beginning and 30.026 m at the end of the work. The area of plan drawn to a scale 1:1000 was found to be 5625 mm^2. Compute the correct area of the field. [**Ans.** 5631.752 m^2]
13. A steel tape 30 m long, standardised at 18°C at a pull of 100 N was used. Find the correction per tape length, if the temperature at the time of measurement is 24°C and the pull exerted is 140 N. Cross-sectional area of the tape is 8 mm^2. Take $E = 2 \times 10^5 N/mm^2$ and $\alpha = 11.2 \times 10^{-6}$ /°C. Tape was stretched over 4 equal spans. If measured length is 200 m, what is the actual length? Take unit weight of steel = 78.6 kN/m^3. [**Ans.** L = 199.979 m]
14. With neat sketches indicate conventional symbol used for the following objects in surveying:
   (*i*) Building (*ii*) Double line railway track
   (*iii*) Cultivated land (*iv*) Tample, mosque and church
   (*v*) River.

4

Compass Surveying

Disadvantage of chain surveying is that, in it only distances are measured and hence area is to be covered with a network of triangles. If the length as well as angle of a line can be measured with respect to a known direction then it is possible to plot a line, independent of length of other lines. Hence, in such cases there is no compulsion of going for a network of triangles only. Compass is an instrument which can be used to measure the direction of a survey line with respect to magnetic north-south. The magnetic north-south direction which is the reference direction is called **meridian** (reference direction) and the angle between the line and the meridian is called **bearing**. Use of compass for measuring direction of a line simplities the surveying to a great extent.

In this chapter construction of different types of compasses, the system of noting bearings of the lines, some problems associated with measurement with compass are explained and then field work involved in compass survey is presented.

4.1 TYPES OF COMPASS

The types of compass that are used commonly are (*i*) prismatic compass and (*ii*) surveyor compass. The essential parts of both type are

(*i*) a magnetic needle

(*ii*) a graduated circle

(*iii*) a line of sight and

(*iv*) a box to house them.

There are some differences in the essential parts of the two type of compass. The construction of the two types of compass is explained and the difference in them is pointed out in this article.

4.1.1 Prismatic Compass

Figure 4.1 shows the cross-section of a typical prismatic compass [see plate 4.1 also].

A magnetic needle of broad form (1) is balanced on a hard and pointed steel pivot (2). The top of the pointed pivot is protected with agate cap (3).

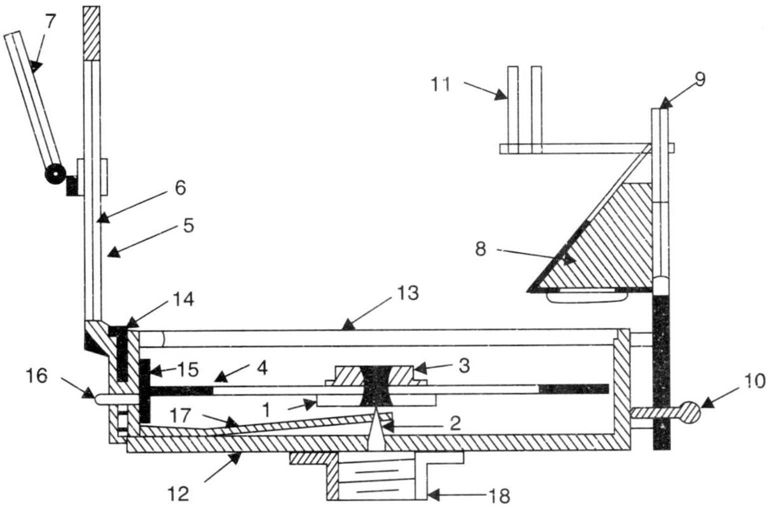

1. Needle
2. Pivot
3. Agate cap
4. Graduated disc
5. Slit metal frame
6. Horse hair
7. Mirror
8. Reflecting prism with cap
9. Eye vane
10. Focussing stud
11. Dark sunglasses
12. Box
13. Glass cover
14. Lifting pin
15. Light spring
16. Brake pin or knob
17. Lifting lever
18. Support to fit on tripod

Fig. 4.1 Prismatic compass

Plate 4.1 Prismatic compass

An aluminium graduated disk (4) is fixed to the top of the needle. The graduations are from zero to 360° in clockwise direction when read from top. The direction of north is treated as zero degrees, east as 90°, south as 180° and west as 270°. However, while taking the readings observations are at the other end of line of sight. Hence, the readings are shifted by 180° and graduations are marked as shown in Fig. 4.2. The graduations are marked inverted because they are read through a prism.

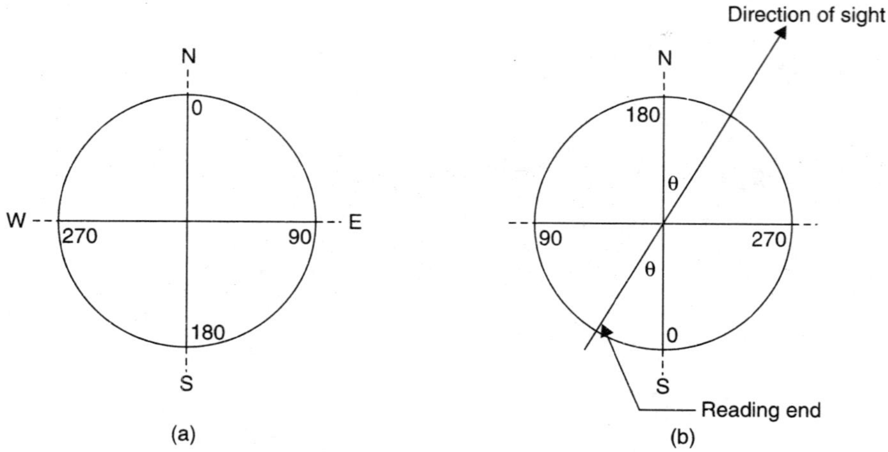

Fig. 4.2

The line of sight consists of object unit and the reading unit. Object unit consists of a slit metal frame (5) hinged to the box. In the centre the slit is provided with a horse hair or a fine wire or thread (6). The metal frame is provided with a hinged mirror (7), which can be placed upward or downward on the frame. It can be slided along the frame. The mirror can be adjusted to view objects too high or too low from the position of compass. Reading unit is provided at diametrically opposite edge. It consists of a prism (8) with a sighting eye vane (9). The prism magnifies the readings on the graduation disk just below it. For focussing, the prism is lowered or raised on the frame carrying it and then fixed with the stud (10). Dark sunglasses (11) provided near the line of sight can be interposed if the object to be sighted is bright (*e.g.*, sun).

The bottom of the box (12) which is about 85 mm to 110 mm supports the pivot of needle firmly at its centre. The object vane and the prism are supported on the sides of the box. The box is provided with a glass (13) lid which protects the graduation disc at the same time permit the direct reading from the top. When the object vane is folded on the glass top it presses a lifting pin (14) which activates lifting lever (15) lifts the needle off the pivot. Thus, it prevents undue wear of pivot point. While taking reading, if graduation disc vibrates, it can be dampened with a spring (16). For pressing spring a knob or brake pin (17) is provided on the box. When not in use prism can be folded over the edge of the box. The box is provided with a lid to close it when the compass is not in use. The box is provided with a socket to fit it on the top of a tripod.

4.1.2 Surveyors Compass

In this type of compass graduation disc is fixed to the box and magnetic needle is free to rotate above it. There is no prism provided at viewing end, but has a narrow slit. After fixing the line of sight, the reading is directly taken from the top of the glass cover. Hence, graduations are written directly (not inverted). In this compass graduations are from zero to 90°, zero being to north or south and 90° being to east and west. An angle of 20° to north direction to the east is written as N 20° E, and an angle of 40° to east from south is written as S 40° E. Always first direction indicated is north or south and the last letter indicates east or west direction. In this system graduated circle rotates with line of sight and magnetic needle is always towards north. The reading is taken at the tip of needle. Hence, on the compass east and west are marked interchanged and marked [Ref. Fig. 4.3] Plate 4.2 shows the photograph of a surveyors compass.

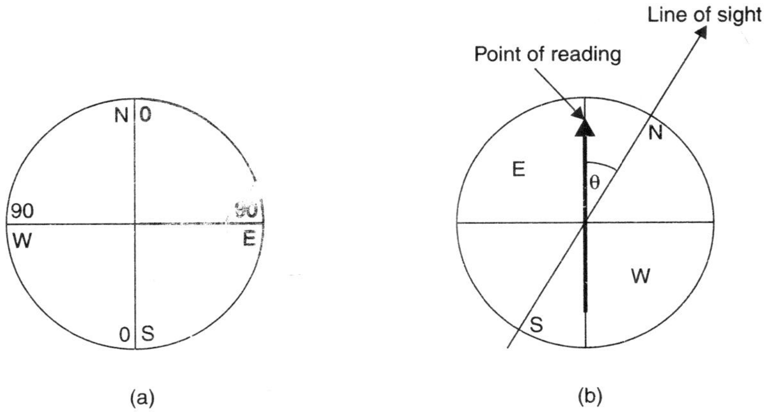

Fig. 4.3

Plate 4.2 Surveyor's compass

Difference Between Prismatic Compass and Surveyors Compass:

The difference between prismatic and surveyor's compass are listed in Table 4.1.

Table 4.1 Differences between prismatic and surveyors compass

Sr. No.	*Prismatic compass*	*Surveyors compass*
1.	Graduation circle is fixed to broad type needle. Hence, it will not rotate with the line of sight.	Graduation circle is fixed to the box. Hence, it rotates with the line of sight.
2.	There is a prism at viewing end.	At viewing end there is no prism. There is only a slit.
3.	Sighting and reading can be done simultaneously.	Sighting and viewing cannot be done simultaneously.
4.	The magnetic needle do not act as an index.	Magnetic needle acts as index while reading.
5.	The graduations are in whole circle bearing.	The graduations are in quadrantal system.
6.	Graduations are marked inverted since its reflection is read through prism.	Graduations are marked directly. They are not inverted.
7.	The reading is taken through a prism.	The reading is taken by directly viewing from top glass.
8.	Tripod may or may not be used. It can be held on a stretched hand also.	Tripod is essential for using it.

4.2 METHOD OF USING A COMPASS

To take a reading from a compass, the following *temporary adjustments* are required:

(*i*) *Centring:* The compass should be fixed to the stand and set over the station. To centre the compass legs of the tripod stand should be moved inward-outward or in a circumferential direction. To check centing plumb may be used or a pebble dropped from the centre of the compass.

(*ii*) *Levelling:* In compass survey perfect levelling is not necessary, but it should be sufficient to permit free suspension of magnetic needle. For checking levelling a bubble level is provided in many compasses. After centring bubble should be ensured in the middle of the circle provided for it in the level. If it is not within that circle, circumferential movements may be provided to the legs of tripod so that without disturbing centring the levelling is achieved.

(*iii*) Focussing the prism: In prismatic compass, to focus the prism on graduated circle, its attachment is slided up or down till the readings are clearly visible. There is no such requirement in surveyors compass.

The following steps are required for observing bearing of a line, say, *AB*.

(*i*) Centre the compass over *A*.

(*ii*) Level the compass.

(*iii*) Focus the prism, if prismatic compass is used.

(*iv*) Rotate the box till ranging rod at *B* is sighted through the line of sight.

(*v*) Bring the needle to rest using knob.

(*vi*) Take the reading and note it in the field book.

Care should be taken to see that the line of sight is not disturbed between the line of sighting the object and the time of reading the bearing.

4.3 BEARING

As stated earlier a bearing of a line is the angle made by the line with respect to a reference direction, the reference direction being known as meridian. The direction shown by a freely suspended and properly balanced magnetic needle is called **magnetic meridian** and the horizontal angle made by a line with this meridian is known as **magnetic bearing.**

The points of intersection of earth's axis with surface of the earth are known as geographic north and south pole. The line passing through geographic north, south and the point on earth is called true **meridian** at that point and the angle made by a line passing through that point is called **true bearing**.

While traversing along lines *A*, *B*, *C*, *D* ..., the bearing of lime *AB* is called **fore bearing of AB** and the bearing of *BA* is called **back bearing**. Fore bearing and back bearing differ by 180°.

4.4 WHOLE CIRCLE BEARING AND REDUCED BEARING

In whole circle bearing (**WCB**) the bearing of a line at any point is measured with respect to a meridian. Its value varies from zero to 360°, increasing in clockwise direction. Zero is north direction, 90° is east,

180° is south and 270° is west (Ref. Fig. 4.2). This type of bearing is used in prismatic compass.

In reduced bearing (**RB**) system, bearings are measured from north or south direction towards east or west. Hence, angles are from 0 to 90° as shown in Fig. 4.3. This system of measuring bearings is used in Surveyor's compass and it is also known as Quadrantal Bearing (**QB**). The bearing measured is designated with letter *N* or *S* in the beginning to indicate whether it is from north or south. The letter *E* or *W* written after the angle indicates whether the bearing read is towards east or west, respectively.

The conversion of the bearing from one system to the other system can be easily carried out by drawing a sketch to indicate *WCB* or *RB* as shown in Fig. 4.4. It may be observed that conversion table is as given below:

Quadrant in which bearing lies	*Conversion relation*
NE	$\alpha = \theta$
SE	$\alpha = 180° - \theta$
SW	$\alpha = \theta - 180°$
NW	$\alpha = 360° - \theta$

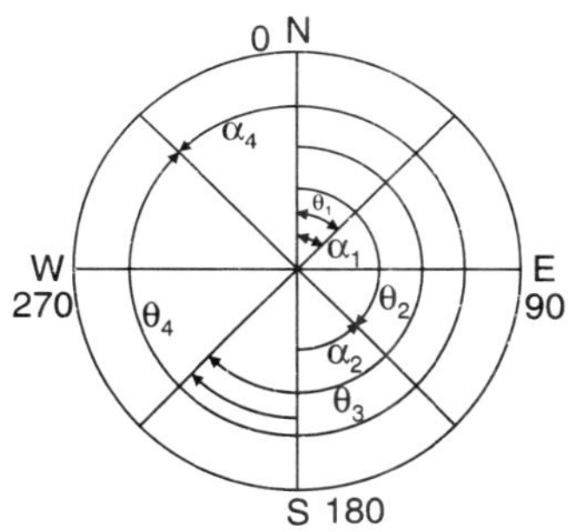

Fig. 4.4

■ **Example 4.1:** *Convert the following reduced bearings into whole circle bearings:*

(*i*) *N 65° E* (*ii*) *S 43° 15′ E*

(*iii*) *S 52° 30′ W* (*iv*) *N 32° 42′ W*

Solution: Let 'θ' be whole circle bearing.

(*i*) Since it is in *NE* quadrant,

$$\theta = \alpha = \mathbf{65°}$$ **Ans.**

(*ii*) Since it is in south east quadrant

$$43° \ 15' = 180° - \theta$$

or $$\theta = 180° - 43° \ 15' = \mathbf{136° \ 45'}$$ **Ans.**

(*iii*) Since it is in SW quadrant

$$52° \ 30' = \theta - 180°$$

or $$\theta = 180° + 52° \ 30' = \mathbf{232° \ 30'}$$ **Ans.**

(*iv*) Since it is in NW quadrant,

$$32° 42' = 360° - \theta$$

or $$\theta = 360° - 32° 42' = \mathbf{327° 18'}$$ **Ans.**

■ **Example 4.2:** *The following fore bearings were observed for lines, AB, BC, CD, DE, EF and FG respectively. Determine their back bearings:*

(*i*) *148°* (*ii*) *65°*

(*iii*) *285°* (*iv*) *215°*

(*v*) *N 36° W* (*vi*) *S 40° E*

Solution: The difference between fore bearing and the back bearing of a line must be 180°. Noting that in WCB angle is from 0° to 360°, we find back bearing = fore bearing ± 180°

+ 180° is used if θ is less than 180° and

– 180° is used when θ is more than 180°.

Hence

(*i*) BB of AB = 145° + 180° = 325°

(*ii*) BB of BC = 65° + 180° = 245°

(*iii*) BB of CD = 285° – 180° = 105°

(*iv*) BB of DE = 215° – 180° = 35°

In case of RB, back bearing of a line can be obtained by interchanging N and S at the same time E and W. Thus

(*v*) BB of EF = S 36° E

(*vi*) BB of FG = N 40° W.

4.5 COMPUTATION OF ANGLES

At any point, if bearings of any two lines are known, the angle between these two lines can be easily found by drawing a neat sketch, and then noting the difference. The procedure is illustrated with the examples given below.

■ **Example 4.3:** *In a closed traverse the following bearings were observed with a compass. Calculate the interior angles.*

Line	*Fore bearing*
AB	*65° 00′*
BC	*125° 30′*
CD	*200° 00′*
DE	*265° 15′*
EA	*330° 00′*

Solution: Figure 4.5 shows the angles observed. Back bearings of all the lines may be worked out and noted on the figure. Then calculation of interior angles can be easily carried out.

Line	*Fore bearing*	*Back bearing*
AB	65° 00′	245° 00′
BC	125° 30′	305° 30′
CD	200° 00′	20° 00′
DE	265° 15′	85° 15′
EA	330° 00′	150° 00′

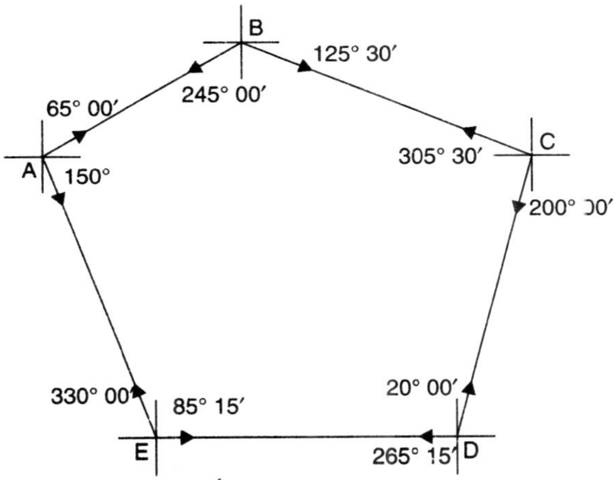

Fig. 4.5

Referring to figure:

$$\angle\mathbf{A} = 150°\ 00' - 65°\ 00' = \mathbf{85°\ 00'}$$

$$\angle\mathbf{B} = 245°\ 00' - 125°\ 30' = \mathbf{119°\ 30'}$$

$$\angle\mathbf{C} = 305°\ 30' - 200°\ 00' = \mathbf{105°\ 30'}$$

$$\angle\mathbf{D} = (360° - 265°\ 15') + 20°\ 00' = \mathbf{114°\ 45'}$$

$$\angle\mathbf{E} = (360° - 330°\ 00') + 85°\ 15' = \mathbf{115°\ 15'}$$

■ **Example 4.4.** *The angles observed with a surveyor compass in traversing the lines AB, BC, CD, DE and EF are as given below. Compute the included angle and show them in a neat sketch.*

Line	*Fore bearing*
AB	*N 55° 30′ E*
BC	*S 63° 30′ E*
CD	*N 70° 00′ E*
DE	*S 48° 30′ E*
EF	*N 72° 15′ E*

Solution: Figure 4.6 shows this traverse. First back bearings of all lines are calculated and noted in the table shown below:

Line	*FB*	*BB*
AB	N 55° 30′ E	S 55° 30′ W
BC	S 63° 30′ E	N 63° 30′ W
CD	N 70° 00′ E	S 70° 00′ W
DE	S 48° 30′ E	N 48° 30′ W
EF	N 72° 15′ E	S 72° 15′ W.

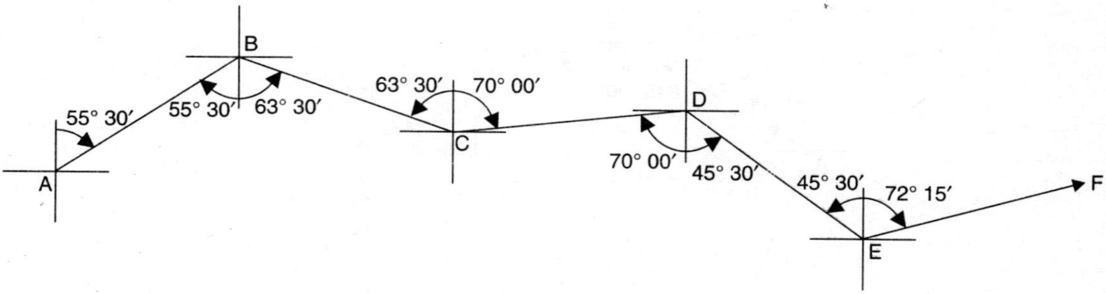

Fig. 4.6

Referring to the figure, we find

$\angle$**B** = 55° 30′ + 63° 30′ = **119° 00′** **Ans.**

$\angle$**C** = 63° 30′ + 70° 00′ = **133° 30′** **Ans.**

$\angle$**D** = 70° 00′ + 45° 30′ = **115° 30′** **Ans.**

$\angle$**E** = 45° 30′ + 72° 15′ = **117° 45′** **Ans.**

4.6 DECLINATION AND DIP

The magnetic meridian and the true meridian may not coincide with each other in a place. The horizontal angle between these two meridians is known as **magnetic declination**. The magnetic north at a place may be towards east or west of true north (Fig. 4.7). If it is towards east, it is known as eastern or +ve declination. Western declination is known as –ve declination. Eastern declination is to be added to observed magnetic bearings to get true meridian. To find magnetic declination at a point true meridian should be established from astronomical observations and magnetic meridian by a compass. Maps are made with respect to true meridian.

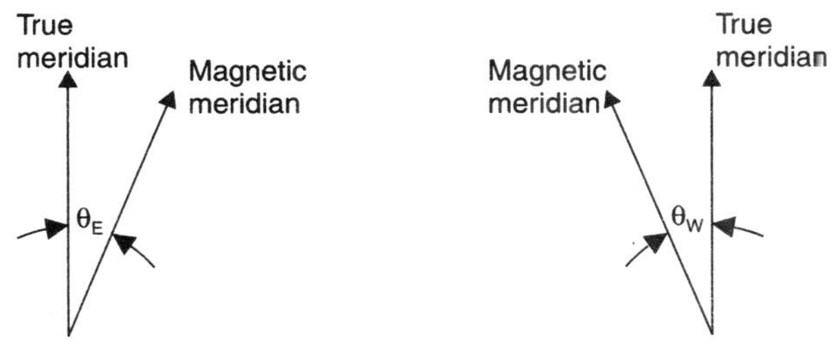

(a) +ve Declination (b) –ve Declination

Fig. 4.7 Magnetic declination

Magnetic declination varies from time to time and also from place to place. In the noon sun is exactly on the geographical meridian. In India, 'Survey of India' department conducts astronomical survey and publishes **Isogonic Charts** from which magnetic declinations at any point can be found. The lines joining the points at which declination is the same at the given time are called **'Isogonic Lines'**. Lines joining points of zero declinations are called '**Agonic Lines**'. The isogonic lines are quite irregular near geographic poles. The isogonic charts show lines of equal annual change in declination.

The following type of variations are observed in declination:

(*i*) Secular variation
(*ii*) Annual variation
(*iii*) Daily variations and
(*iv*) Irregular variations.

4.6.1 Secular Variation

The magnetic meridian swings like a pendulum to the left and to the right of true meridian. Its period of variation is approximately 250 years.

4.6.2 Annual Variation

It is observed that in a year declination varies from $1'$ to $2'$.

4.6.3 Daily Variation

The daily variation of magnetic declination is as much as $10'$. This variation is also known as '**Dirunal Variation**'. The following factors influence its magnitude:

(*a*) It is more in day and less in night.
(*b*) It is more in summer and less in winter.
(*c*) The amount of variation changes from year to year.
(*d*) It is more near magnetic poles and less near equator.

4.6.4 Irregular Variation

Due to earthquakes and volcanic eruptions, magnetic storms occur, resulting into changes in magnetic meridian. Such changes are from $1°$ to $2°$.

Magnetic Dip

A perfectly balanced, freely suspended magnetic needle dips towards its northern end in northern hemisphere and towards its southern end in southern hemisphere. If it is at north pole, the needle takes vertical position. The vertical angle between the horizontal and the direction shown by a perfectly balanced and freely suspended needle is known as the magnetic dip at that place. Its value is 0° at equator and 90° at magnetic poles. To counteract the dip, a sliding rider (weight) is provided on the needle.

■ **Example 4.5:** *True bearing of line AB is 357° and its magnetic bearing is 1° 30′. Determine the declination. Also find the true bearing of AC which has magnetic bearing equal to 153° 30′.*

Solution: [Ref. Fig. 4.8]

Magnetic Declination = 1° 30′ + (360° – 357°) = 4° 30′, west. Magnetic bearing of AC = 153° 30′.

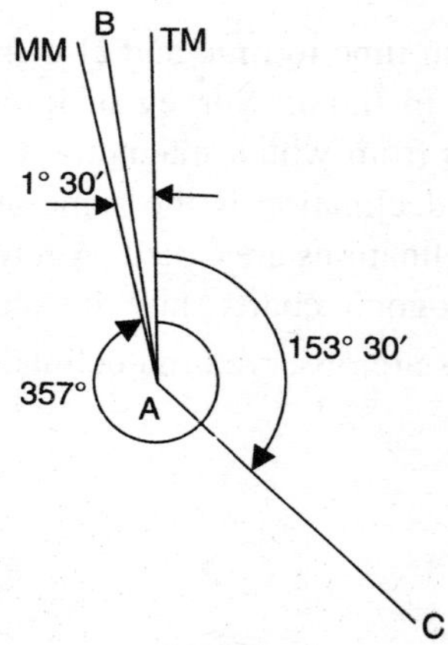

Fig. 4.8

∴ **True bearing of AC** = 153° 30′ – 4° 30′

= **149° Ans.**

■ **Example 4.6:** *In an old map line AB was drawn with a magnetic bearing of 136° 45′ when the magnetic declination was 2° 30′ east. To what magnetic bearing the line should be set now, if magnetic declination is 3° 30′ west ?*

Solution: Referring to Fig. 4.9, it is clear that

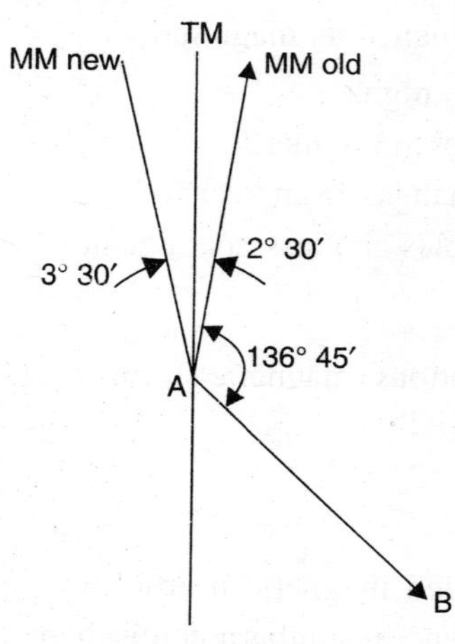

Fig. 4.9

Present magnetic bearing of line AB

$$= 3° 30' + 2° 30' + 136° 45'$$

$$= \mathbf{142° 45'}$$ **Ans.**

■ **Example 4.7:** *Find the declination at a place, if the magnetic bearing to the sun at noon is (a) 185° (b) 358°.*

Solution: (*a*) Referring to Fig. 4.10 (*a*), since the magnetic bearing to the sun in the noon is 185°, it is to the geographic south pole. Hence, to the magnetic bearing to north pole is 5°, *i e.*, **declination is 5°. Ans.**

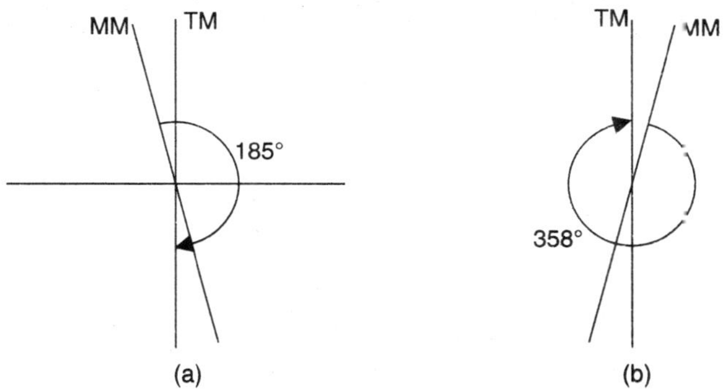

Fig. 4.10

(*b*) Referring to Fig. 4.10 (*b*),

Declination = 360° – 358° = **2° East.**

4.7 LOCAL ATTRACTION

A freely suspended and properly balanced magnetic needle is expected to show magnetic meridian. However, local objects like electric wires and objects of steel attract magnetic needle towards themselves. Thus, needle is forced to show slightly different direction. This disturbance is called local attraction. The list of materials which cause local attraction are:

(*i*) magnetic rock or iron ore,

(*ii*) steel structures, iron poles, rails, electric poles and wires,

(*iii*) key bunch, knife, iron buttons, steel rimmed spectacles, and

(*iv*) chain, arrows, hammer, clearing axe etc.

Surveyor is expected to take care to avoid local attractions listed in (*iii*) and (*iv*) above.

Detecting Local Attraction

For detecting local attraction it is necessary to take both fore bearing and back bearing for each line. If the difference is exactly 180°, the two stations may be considered as not affected by local attraction. If difference is not 180°, better to go back to the previous station and check the fore bearing. If that reading is same as earlier, it may be concluded that there is local attraction at one or both stations.

Correcting Observed Bearings

If local attraction is detected in a compass survey observed bearings may be corrected by any one of the following two methods:

Method I: It may be noted that the included angle is not influenced by local attraction as both readings are equally affected. Hence, first calculate included angles at each station, commencing from the unaffected line and using included angles, the corrected bearings of all lines may be calculated.

Method II: In this method, errors due to local attraction at each of the affected station is found starting from the bearing of a unaffected local attraction, the bearing of the successive lines are adjusted.

These methods are illustrated with the example below:

■ **Example 4.8:** *In a closed traverse, the following bearings were observed, with a compass. Calculate their interior angles and then compute the corrected magnetic bearings:*

Line	*FB*	*BB*
AB	*46° 30′*	*226° 30′*
BC	*118° 30′*	*300′ 15′*
CD	*210° 00′*	*28° 00′*
DE	*271° 15′*	*93° 15°*
EA	*313° 45′*	*131° ...*

Solution: Figure 4.11 shows the traverse. Difference between FB and BB of line AB is exactly 180°. Hence, stations A and B are not affected by local attractions. Hence, all the bearings taken from stations A and B are correct magnetic bearings.

Correct bearing of AB = 46° 30′

Correct bearing of BA = 226° 30′

From the figure,

$$\angle A = 132° \; 00' - 46° \; 30' = 85° \; 30'$$

$$\angle B = 226° \; 30' - 118° \; 30' = 108° \; 00'$$

$$\angle C = 300° \; 15' - 210° \; 00' = 90° \; 15'$$

$$\angle D = (360° - 271° \; 15') + 28° \; 00' = 116° \; 45'$$

$$\angle E = (360° \; 00' - 313° \; 45') + 93° \; 15' = 139° \; 30'$$

Total Interior Angle $= \angle A + \angle B + \angle C + \angle D + \angle E$

$= 540° \; 00'$.

Hence there is no observations error.

[*Note:* In a pentagon sum of interior angles = $(2n - 4) \times 90 = (2 \times 5 - 4) \times 90 = 540°$. If there is observation error it is to be distributed equally to all interior angles].

Since, stations A and B are not affected by local attraction, correct bearings are:

Bearing of AB = 46° 30′

Bearing of BA = 46° 30′ + 180° 00′ = 226° 30′

Bearing of BC = 226° 30′ – ∠B = 226° 30′ – 108° 00′ = 118° 30′

Bearing of CB = 118° 30′ + 180° 00′ = 298° 30′

Bearing of CD = 298° 30′ – ∠C = 298° 30′ – 90° 15′ = 208° 15′

Bearing of DC = 208° 15′ – 180° 00′ = 28° 15′

Bearing of DE = 28° 15′ – ∠D = 28° 15′ – 116° 45′

= – 89° 30′ = – 88° 30′ + 360° 00′ = 271° 30′

Bearing of ED = 271° 30′ – 180° 00′ = 91° 30′

Bearing of EA = 91° 30′ – ∠E = 90° 30′ – 139° 30′

= – 48° 00′ = – 48° 00′ + 360° = 312° 00′

Bearing of AE = 312° 00′ – 180° 00′ = 132° 00′

[Checked. It should be equal to the observed bearing, since station E is not affected].

■ **Example 4.11:** *Solve problem no. 10 by the method of correction to local attraction.*

Solution: [Ref. Fig. 4.11]

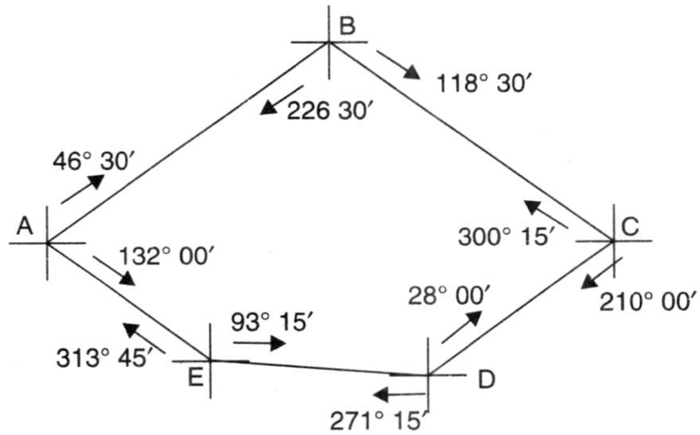

Fig. 4.11

Since, the difference between FB and BB of line AB is exactly 180°, stations A and B are not affected by local attraction. Hence, corrections to the observed bearings at A and B are zero.

∴ Correct bearing CB = 118° 30′ + 180° 00′ = 298° 30′

But observed bearing = 300° 15′

Hence correction at station C = 298° 30′ – 300° 15′ = – 1° 45′

∴ Correct bearing of CD = 210° 00′ – 1° 45′ = 208° 15′

Correct bearing DC = 208° 15′ – 180° 00′ = 28° 15′

Observed bearing of DC = 28° 00′

∴ Corrections required at D = 28° 18′ – 28° 00′ = 0° 15′

∴ Correct bearing of DE = 271° 15′ + 0° 15′ = 271° 30′

Correct bearing of ED = 271° 30′ – 180° 00′ = 91° 30′

But observed bearing of ED = 93° 15′

∴ Correction for observations at E = 91° 30′ – 93° 15′ = – 1° 45′

∴ Correct bearing of EA = 313° 45′ – 1° 45′ = 312° 00′

∴ Correct bearing of AE = 312° 00′ – 180° 00′ = 132° 00′. **[Checked]**

The calculations may be carried out in tabular form as shown in Table 4.2.

Table 4.2

Station	*Line*	*Observed bearing*	*Correction*	*Correct bearing*
A	AE AB	132° 00′ 46° 30′	0	132° 00′ 46° 30′
B	BA BC	226° 30′ 118° 30′	0	226° 30′ 118° 30′
C	CB CD	300° 15′ 210° 00′	– 1° 45′	300° 15′ 210° 00′
D	DC DE	28° 00′ 271° 15′	0° 15′	28° 00′ 271° 15′
E	ED EA	93° 15′ 313° 45′	– 1° 45′	93° 15′ 313° 45′

[*Note:* Correction = Correct reading – Observed reading]

4.8 CHAIN AND COMPASS SURVEYING FIELD WORK

In compass survey chain or tape is used for linear measurement. If the surveying starts from a station, goes round an area and ends at the starting station it is called **closed traverse.** If survey starts from a point, goes along a number of interconnected lines and ends at some other point it is called as **open traverse.** Closed traverse is used for preparing plan of an area while open traverse is useful in the road, rail or canal projects. The following are required for chain and compass survey:

(*i*) Compass and stand

(*ii*) Chain and tape

(*iii*) 10 arrows

(*iv*) 5 to 6 ranging rods

(*v*) Ranging poles

(*vi*) Pegs and hammer

(*vii*) Plumb bobs

(*viii*) Line ranger and cross staff etc.

Field Work

Field work involves:

(*i*) reconnaissance survey
(*ii*) preparation of location sketches of stations
(*iii*) measurement of directions
(*iv*) measurement of lengths and offsets, and
(*v*) recording measurements.

(*i*) **Reconnaissance Survey:** The entire area to be surveyed is inspected to select survey stations. Important points to be considered in selecting stations are:

(*a*) Adjacent stations should be intervisible.
(*b*) Chaining the liners should be free of obstacles.
(*c*) Number of survey lines should be minimum.
(*d*) Survey lines should run close to the important objects, so that offset lengths are small.

An index map is prepared with pencil and stations are marked. If necessary, changes may be made in survey lines and corresponding changes in the index plan.

Location Sketches

Before commencing surveying a line, the location sketches of the stations of that line should be prepared. At the beginning of the field book few pages should be reserved for drawing location sketches.

Direction Measurement

The following precautions should be taken in measuring the direction of a survey line with compass.

(*a*) Centre the compass on the station correctly.
(*b*) Level the compass and ensure needle is free to move.
(*c*) Take the reading only after vibration of graduation circle/needle is stopped. Use the knob, if necessary.
(*d*) Gently tap the top of the glass of compass to remove sluggishness of the needle and take the reading after vibration stops.
(*e*) While taking reading parallax should be avoided.
(*f*) Care should be taken to keep away steel and iron objects like key bunch, metal framed objects, iron buttons, chain, arrows etc.
(*g*) If handkerchief is used to clean top of glass of campass, the glass is charged with electricity. As a result of it local attraction is induced. To avoid this problem apply moist fingers to clean the glass.
(*h*) If the compass is not in use, fold the prism and object vane on the top of glass plate, so that needle is lifted from the pivot to avoid unnecessary wear of the pivot.
(*i*) For all survey lines fore bearings and back bearings should be taken. If any other survey station is visible, bearing should be taken to that station also, which helps in checking survey work.

Measurement of Lengths and Offsets

This is similar to the one used in chain survey. However, it may be noted that for the objects of less importance one can measure the bearing of the offset and its length.

Recording Chain and Compass Measurements

The type of field book used in chain survey is used in this survey also. Apart from recording linear measurements in this survey the bearings taken also should be recorded. Figure 4.12 shows a page typical field record.

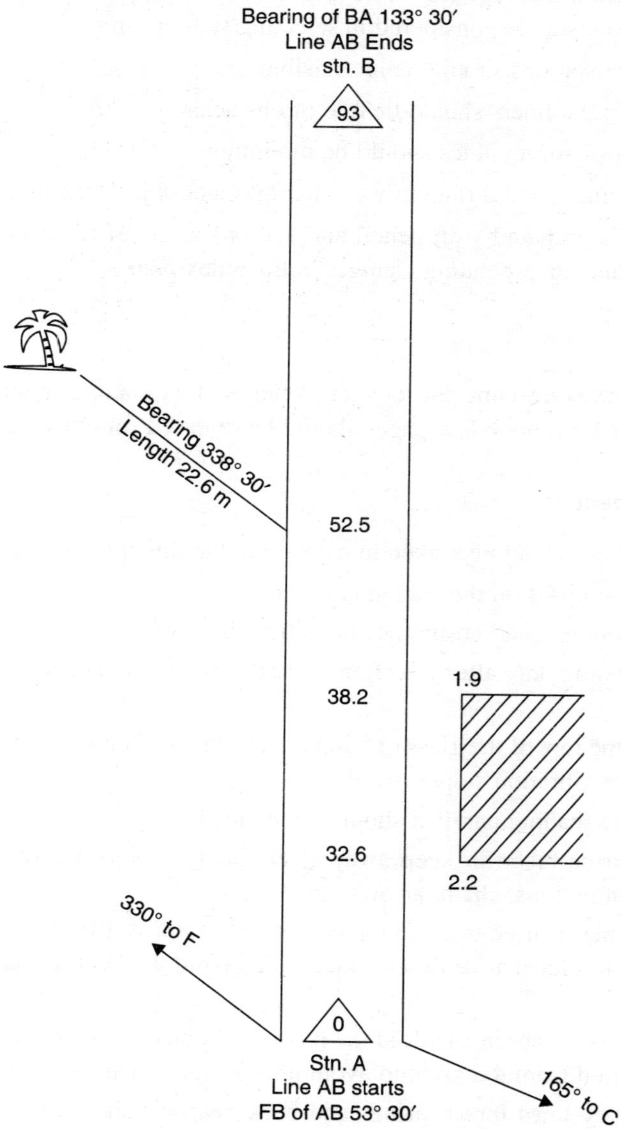

Fig. 4.12

QUESTIONS

1. Explain with a neat sketch construction of a prismatic compass.
2. Bring out the differences between prismatic compass and surveyor's compass.
3. Distinguish between
   (*a*) Magnetic meridian and true meridian.
   (*b*) Whole circle bearing and quadrantal bearing.
   (*c*) Declination and dip.
   (*d*) Fore bearing and back bearing.
   (*e*) Isogonic and agonic lines.
4. What is meant by magnetic declination? List the different types of its variations.
5. What is local attractions? How it is detected in the field?
6. What are the precautions to be taken while taking bearing of a line with a compass.
7. In traversing in anticlockwise direction, the following readings were observed:

Line	AB	BC	CD	DE	EA
FB	105° 15′	20° 00′	316° 30′	187° 15′	122° 45′

Draw a neat sketch of the traverse. Determine the interior angles of the traverse and apply check.

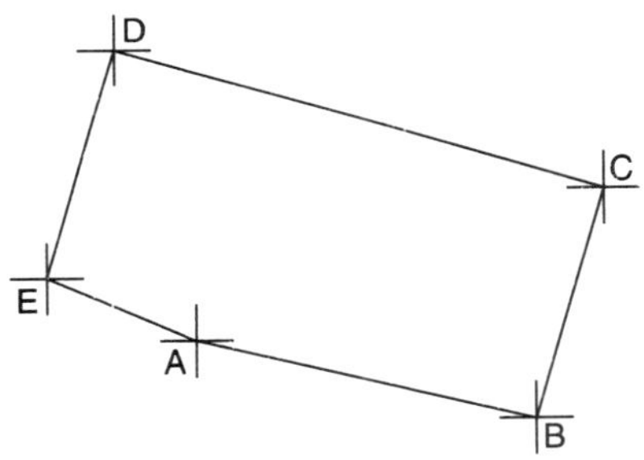

∠A = 162° 30′ ∠B = 94° 45′

∠C = 116° 30′ ∠D = 50° 45′

∠E = 115° 30′ Σθ = 540° Checked. **Ans.**

8. The following bearings were taken in running a compass traverse:

Line	*FB*	*BB*
AB	124° 30′	304° 30′
BC	68° 15′	246° 00′
CD	310° 30′	135° 15′
DA	200° 15′	17° 45′

At what stations do you suspect local attraction? Find the correct bearings of the lines and also compute the included angles.

Ans. Stations C and D are affected by local attraction. Correct bearing of lines are as shown below:

Line	*FB*	*BB*
AB	124° 30′	304° 30′
BC	68° 15′	248° 15′
CD	312° 45′	132° 45′
DA	197° 45′	17° 45′

$\angle A = 106° 45'$ $\angle B = 123° 45'$ $\angle C = 64° 30'$ $\angle D = 65°$ $\Sigma\theta = 360°$ **Checked.**

5

Elevation Measurements

Elevation measurements involve measurements in vertical plane. It is also known as levelling. It may be *defined as the art of determining the elevations of given points above or below a datum line or establishing given points of required heights above or below the datum line.*

5.1 OBJECT AND USES OF LEVELLING

As stated in the definition of levelling, the object is

(*i*) to determine the elevations of given points with respect to a datum

(*ii*) to establish the points of required height above or below the datum line.

Uses of levelling are

(*i*) to determine or to set the plinth level of a building.

(*ii*) to decide or set the road, railway, canal or sewage line alignment.

(*iii*) to determine or to set various levels of dams, towers, etc.

(*iv*) to determine the capacity of a reservoir.

5.2 TERMS USED IN LEVELLING

Before studying the art of levelling, it is necessary to clearly understand the following terms used in levelling:

1. **Level Surface:** A surface parallel to the mean spheroid of the earth is called a level surface and the line drawn on the level surface is known as a level line. Hence all points lying on a level surface are equidistant from the centre of the earth. Figure 5.1 shows a typical level surface.

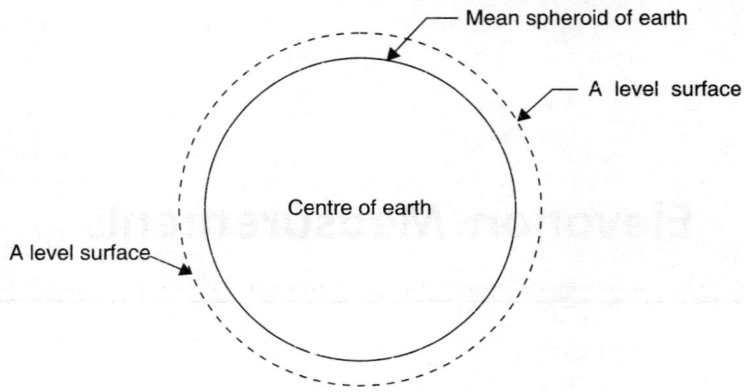

Fig. 5.1 A level surface

2. **Horizontal Surface:** A surface tangential to level surface at a given point is called horizontal surface at that point. Hence a horizontal line is at right angles to the plumb line at that point [Ref. Fig. 5.2].

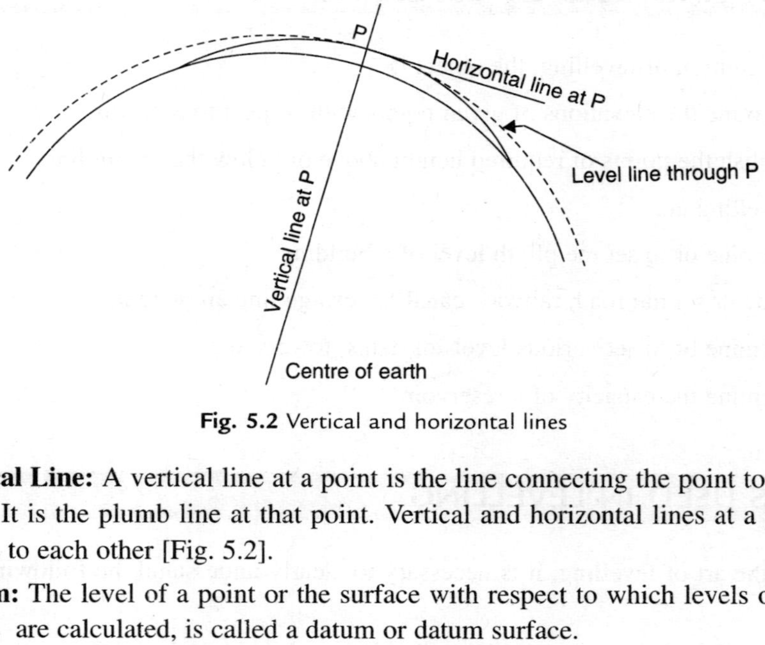

Fig. 5.2 Vertical and horizontal lines

3. **Vertical Line:** A vertical line at a point is the line connecting the point to the centre of the earth. It is the plumb line at that point. Vertical and horizontal lines at a point are at right angles to each other [Fig. 5.2].
4. **Datum:** The level of a point or the surface with respect to which levels of other points or planes are calculated, is called a datum or datum surface.
5. **Mean Sea Level (MSL):** MSL is the average height of the sea for all stages of the tides. At any particular place MSL is established by finding the mean sea level (free of tides) after averaging tide heights over a long period of at least 19 years. In India MSL used is that established at Karachi, presently, in Pakistan. In all important surveys this is used as datum.
6. **Reduced Levels (RL):** The level of a point taken as height above the datum surface is known as RL of that point.

7. **Bench Marks:** A bench mark is a relatively permanent reference point, the elevation of which is known (assumed or known w.r.t. MSL). It is used as a starting point for levelling or as a point upon which to close for a check. The following are the different types of bench marks used in surveying:

(*a*) GTS bench marks
(*b*) Permanent bench marks
(*c*) Arbitrary bench marks and
(*d*) Temporary bench marks.

(*a*) **GTS Bench Mark:** The long form of GTS bench mark is Great Trigonometrical Survey bench mark. These bench marks are established by national agency. In India, the department of Survey of India is entrusted with such works. GTS bench marks are established all over the country with highest precision survey, the datum being mean sea level. A branch plate provided on the top of a concrete pedastal with elevation engraved on it serves as bench mark. It is well protected with masonry structure built around it so that its position is not disturbed by animals or by any unauthorised person. The GTS bench marks are shown in the topo sheets published.

(*b*) **Permanent Bench Mark:** These are the bench marks established by state government agencies like PWD. They are established with reference to GTS bench marks. They are usually on the corner of plinth of public buildings.

(*c*) **Arbitrary Bench Mark:** In many engineering projects the difference in elevations of neighbouring points is more important than their reduced level with respect to mean sea level. In such cases a relatively permanent point, like plinth of a building or corner of a culvert, are taken as bench marks, their level assumed arbitrarily such as 100.0 m, 300.0 m, etc.

(*d*) **Temporary Bench Mark:** This type of bench mark is established at the end of the day's work, so that the next day work may be continued from that point. Such point should be on a permanent object so that next day it is easily identified.

5.3 LEVELLING INSTRUMENTS

A level is an instrument giving horizontal line of sight and magnifying the reading at a far away distance. It consists of the following parts:

(*i*) A telescope to provide a line of sight
(*ii*) A level tube to make the line of sight horizontal and
(*iii*) A levelling head to level the instrument.

The following types of levels are available:

(*i*) Dumpy level
(*ii*) Wye (or, *Y*) level
(*iii*) Cooke's reversible level
(*iv*) Cushing's level
(*v*) Tilting level and
(*vi*) Auto level.

5.3.1 Dumpy Level

It is a short and stout instrument with telescope tube rigidly connected to the vertical spindle. Hence the level tube cannot move in vertical plane. It cannot be removed from its support. Hence it is named as dumpy level. The telescope rotates in horizontal plane in the socket of the levelling head. A bubble tube is attached to the top of the telescope. Figure 5.3 shows a typical dumpy level. Plate 5.1 shows its photograph.

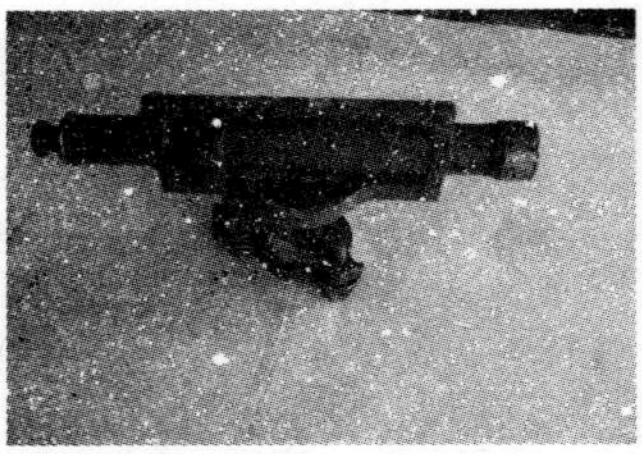

Plate 5.1 Dumpy level

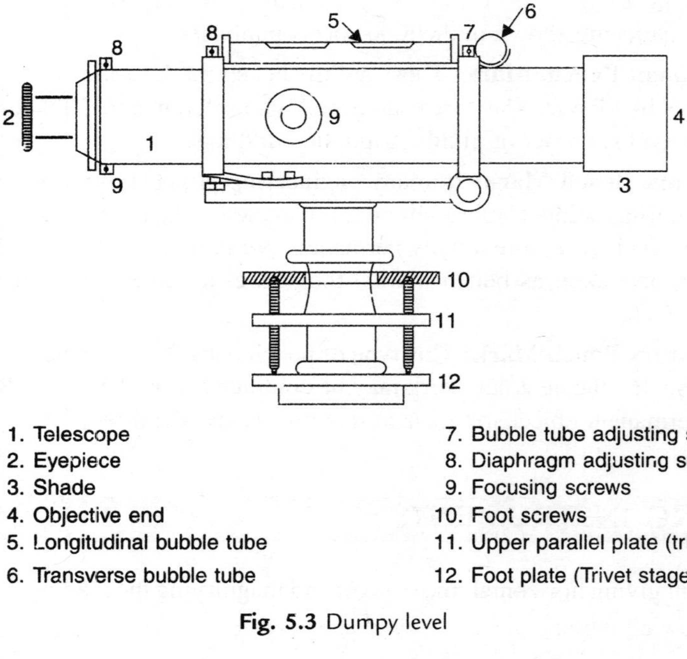

1. Telescope
2. Eyepiece
3. Shade
4. Objective end
5. Longitudinal bubble tube
6. Transverse bubble tube
7. Bubble tube adjusting screws
8. Diaphragm adjusting screws
9. Focusing screws
10. Foot screws
11. Upper parallel plate (tribrach)
12. Foot plate (Trivet stage)

Fig. 5.3 Dumpy level

Telescope is a tube with object glass and eyepiece. Object glass can be adjusted using the focussing screw before sighting the graduated staff held on the object. Eyepiece can be adjusted by rotating it to see that parallel is removed and cross hairs appears distinctly. Eyepiece once adjusted needs no change as long as the same person takes the readings.

Level tube is a glass tube with slightly curved shape provided over the level tube. The tube is filled with ether or alcohol leaving a little air gap, which takes the shape of a bubble. The air bubble is always at the highest point. The level tube is fixed with its axis parallel to telescope tube, so that when bubble is centred, the telescope is horizontal. The tube is graduated on either side of its centre to estimate how much the bubble is out of centre. The glass tube is placed inside a brass tube which is open from top and lower side it is fixed to telescope tube by means of capston headed nuts. The bubble tube is adjusted with these nuts, if it is out of order.

Levelling head consists of two parallel plates with three foot screws. The upper plate is known as tribratch plate and the lower one as the trivet. The lower plate can be screwed on to the tripod stand. By adjusting the screws the instrument can be levelled to get perfect horizontal line of sight.

Dumpy level is to be fitted to a tripod stand to use it in the field. The tripod stand consists of three legs connected to a head to which the lower plate of level can be fitted. The lower side of the legs are provided with metal shoes to get good grip with ground. Plate 5.2 shows typical level stands.

Plate 5.2 Levelling stands (adjustable and non-adjustable)

5.3.2 Wye or Y-Level

In this type of level, the telescope in supported in two *y*-shaped supports and can be fixed with the help of curved clips. Clips can be opened and telescope can be reversed end to end and fitted. The advantage of this level is some of the errors eliminated, if the readings are taken in both the direction of telescope.

5.3.3 Cooke's Reversible Level

In this instrument the telescope is supported by two rigid sockets into which telescope can be introduced from either end and then screwed. For taking the readings in the reversed position of telescope, the screw is slackened and then the telescope is taken out and reversed end for end. Thus it combines the rigidity of dumpy level and reversibility of *y*-level.

5.3.4 Cushings Level

In this reversing of telescope end for end is achieved by interchanging the eyepiece and the objective piece since both collars are exactly the same.

5.3.5 Tilting Level

In this, telescope can be tilted through about *n* degrees with the help of a tilting screw. Hence bubble can be easily centered. But it needs centering of the bubble before taking every reading. Hence it is useful, if at every setting of the instrument a number of readings to be taken are few.

5.3.6 Auto Level

The auto-level or the automatic-level is a self aligning level. Within a certain range of tilt automatic levelling is achieved by an inclination compensating device. The operational comfort, high speed and precision are the advantages of this instrument.

5.4 LEVELLING STAFF

Along with a level, a levelling staff is also required for levelling. The levelling staff is a rectangular rod having graduations. The staff is provided with a metal shoes at its bottom to resist wear and tear. The foot of the shoe represents zero reading. Levelling staff may be divided into two groups:

(*i*) Self reading staff (*ii*) Target staff.

(*i*) **Self reading staff:** This staff reading is directly read by the instrument man through telescope. In a metric system staff, one metre length is divided into 200 subdivisions, each of uniform thickness of 5 mm. All divisions are marked with black in a white background. Metres and decimetres are written in red colour [Fig 5.4 (*a*)]. The following three types of self reading staffs are available:

(*a*) *Solid staff:* It is a single piece of 3 m.

(*b*) *Folding staff:* A staff of two pieces each of 2 m which can be folded one over the other.

(*c*) *Telescopic staff:* A staff of 3 pieces with upper one solid and lower two hollow. The upper part can slide into the central one and the central part can go into the lower part. Each length can be pulled up and held in position by means of brass spring. The total length may be 4 m or 5 m [Fig. 5.4 (*b*)].

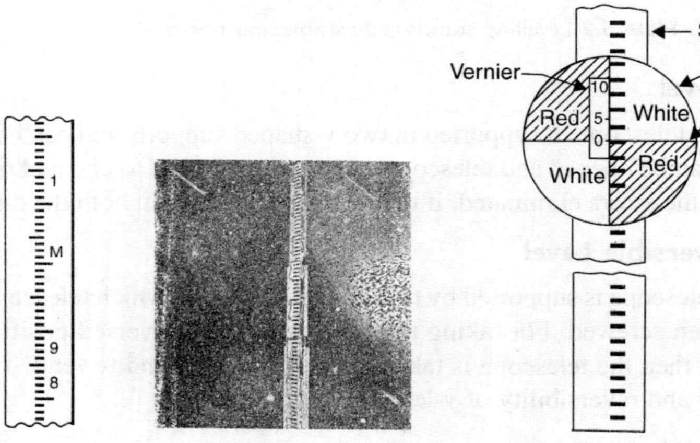

(a) Self-reading staff (b) Telescopic staff (c) Target staff

Fig. 5.4

(*ii*) **Target staff:** If the sighting distance is more, instrument man finds it difficult to read self reading staff. In such case a target staff shown in [Fig. 5.4 (*c*)] may be used. Target staff is similar to self reading staff, but provided with a movable target. Target is a circular or oval shape, painted red and white in alternate quadrant. It is fitted with a vernier at the centre.

The instrument man direct the person holding target staff to move the target, till its centre is in the horizontal line of sight. Then target man reads the target and is recorded.

5.5 METHODS OF LEVELLING

The following methods are used to determine the difference in elevation of various points.

(*i*) Barometric levelling (*ii*) Hypsometric levelling

(*iii*) Direct levelling and (*iv*) Indirect levelling.

5.5.1 Barometric Levelling

This method depends on the principle that atmospheric pressure depends upon the elevation of place. Barometer is used to measure the atmospheric pressure and hence elevation is computed. However it is not accurate method since the atmospheric pressure depends upon season and temperature also. It may be used in exploratory surveys.

5.5.2 Hypsometric Levelling

This is based on the principle that boiling point of water decreases with the elevation of the place. Hence the elevation difference between two points may be found by noting the difference in boiling point of water in the two places. This method is also useful only for exploratory survey.

5.5.3 Direct Levelling

It is common form of levelling in all engineering projects. In this method horizontal sight is taken on a graduated staff and the difference in the elevation of line of sight and ground at which staff is held are found. Knowing the height of line of sight from the instrument station the difference in the elevations of instrument station and the ground on which staff is held can be found. This method is thoroughly explained in next article.

5.5.4 Indirect Methods

In this method instruments are used to measure the vertical angles. Distance between the instrument and staff is measured by various methods. Then using trigonometric relations, the difference in elevation can be computed. This is considered beyond the scope of this book. One can find details of such methods in books on surveying and levelling.

5.6 TERMS USED IN DIRECT METHOD OF LEVELLING

The following terms are used in direct method of levelling:

(*i*) **Plane of Collimation:** It is the reduced level of plane of sight with respect to the datum selected. It is also known as **'height of instrument'**. It should not be confused with the height of telescope from the ground where the instrument is set.

(*ii*) **Back Sight (BS):** It is the sight taken on a level staff held on the point of known elevation with an intension of determining the plane of collimation. It is always the first reading after the instrument is set in a place. It is also known as plus sign, since this reading is to be added to RL of the point (Bench mark or change point) to get plane of collimation.

(*iii*) **Intermediate Sight (IS):** Sights taken on staff after back sight (first sight) and before the last sight (fore sight) are known as intermediate sights. The intension of taking these readings is to find the reduced levels of the points where staff is held. These sights are known as 'minus sights' since the IS reading is to be subtracted from plane of collimation to get R L of the point where staff is held.

(*iv*) **Fore Sight (FS):** This is the last reading taken from the instrument station before shifting it or just before ending the work. This is also a minus sight.

(*v*) **Change Point (CP):** This is also known as turning point (TP). This is a point on which both fore sights and back sights are taken. After taking fore sight on this point instrument is set at some other convenient point and back sight is taken on the staff held at the same point. The two readings help in establishing the new plane of collimation with respect to the earlier datum. Since there is time gap between taking the two sights on the change point, it is advisable to select change point on a well defined point.

5.7 TEMPORARY ADJUSTMENTS OF A LEVEL

The adjustments to be made at every setting of the instrument are called temporary adjustments. The following three adjustments are required for the instrument whenever set over a new point before taking a reading:

(*i*) Setting (*ii*) Levelling and

(*iii*) Focussing.

5.7.1 Setting

Tripod stand is set on the ground firmly so that its top is at a convenient height. Then the level is fixed on its top by turning trivet stage of level. By turning tripod legs radially or circumferentially, the instrument is approximately levelled. Some instruments are provided with a less sensitive circular bubble on tribrach for this purpose.

5.7.2 Levelling

The procedure of accurate levelling with three levelling screw is as given below:

(*i*) Loosen the clamp and turn the telescope until the bubble axis is parallel to the line joining any two screws [Ref. Fig. 5.5 (*a*)].

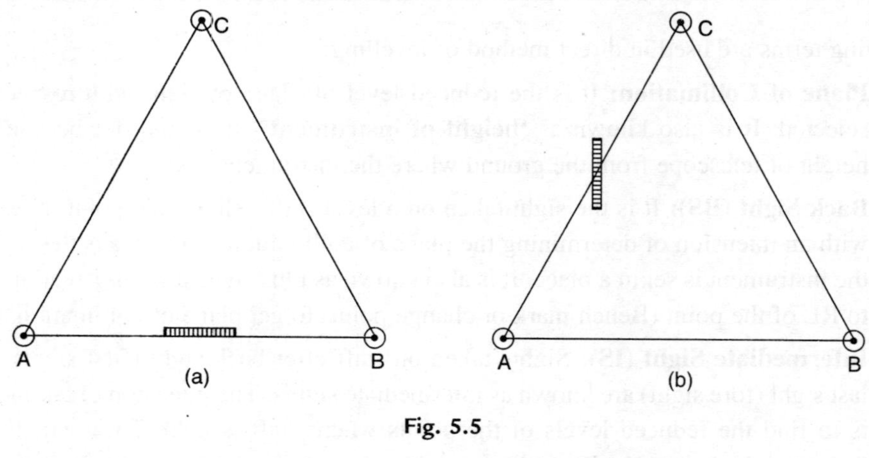

Fig. 5.5

(*ii*) Turn the two screws inward or outward equally and simultaneously till bubble is centred.

(*iii*) Turn the telescope by 90° so that it lies over the third screw [Fig. 5.4 (*b*)] and level the instrument by operating the third screw.

(*iv*) Turn back the telescope to its original position [Fig. 5.5 (*a*)] and check the bubble. Repeat steps (*ii*) to (*iv*) till bubble is centred for both positions of the telescope.

(*v*) Rotate the instrument by 180°. Check the levelling.

5.7.3 Focussing

Focussing is necessary to eliminate parallax while taking reading on the staff. The following two steps are required in focussing:

(*i*) **Focussing the eyepiece:** For this, hold a sheet of white paper in front of telescope and rotate eyepiece in or out till the cross hairs are seen sharp and distinct.

(*ii*) **Focussing the objective:** For this telescope is directed towards the staff and the focussing screw is turned till the reading appears clear and sharp.

5.8 TYPES OF DIRECT LEVELLING

The following are the different types of direct levelling:

(*i*) Simple levelling
(*ii*) Differential levelling
(*iii*) Fly levelling
(*iv*) Profile levelling
(*v*) Cross sectioning and
(*vi*) Reciprocal levelling.

5.8.1 Simple Levelling

It is the method used for finding difference between the levels of two nearby points. Figure 5.6 shows one such case in which level of A is assumed, say 200.00 m. RL of B is required.

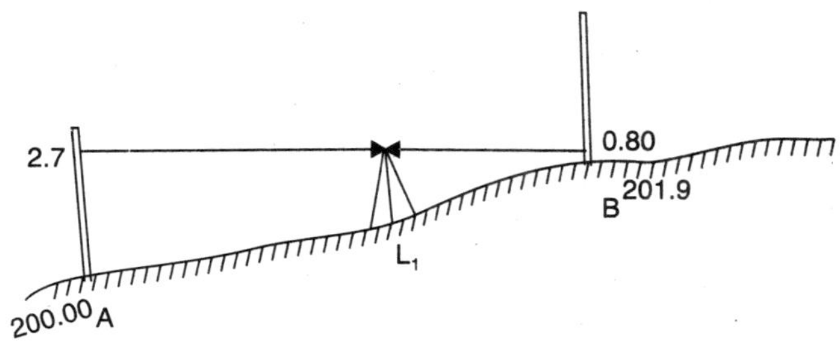

Fig. 5.6

RL of A = 200.00 m

Back sight on A = 2.7 m.

∴ Plane of collimation for setting at station = 200 + 2.7
= 202.7 m

Fore sight on B = 0.80 m

∴ RL of B = 202.7 – 0.80
= 201.9 m

It may be noted that the instrument station L_1 need not be along the line AB (in plan). Since RL of L_1 do not appear in the calculations.

5.8.2 Differential Levelling

If the distance between two points A and B is large, it may not be possible to take the readings on A and B from a single setting.

In such situation differential levelling is used. In differential levelling the instrument is set at more than one position, each shifting facilitated by a change point. Figure 5.7 shows a scheme of such setting.

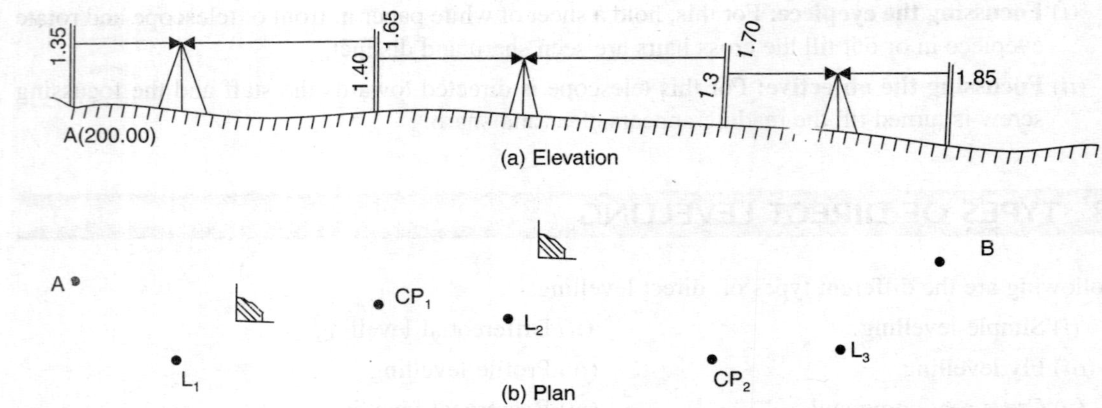

Fig. 5.7

RL of A is 200.00 m. Instrument is set up at L_1 and back sight on A is 1.35 m. The fore sight on change point CP_1 is 1.65 m. Then instrument is shifted to L_2 and back sight on CP_1 is 1.40 m. Fore sight on CP_2 is 1.70 m. After this instrument is shifted to L_3 and back sight on CP_2 is 1.3 m. The work ended with a fore sight of 1.85 m on B. The RL of B is to be found.

RL of A = 200.00 m

Back sight on A = 1.35 m

∴ Plane of collimation at L_1 = 200 + 1.35 = 201.35 m

Fore sight on CP_1 = 1.65 m

∴ RL of CP_1 = 201.35 – 1.65 = 199.70 m

Back sight to CP_1 from L_2 = 1.40

∴ Plane of collimation at L_2 = 199.70 + 1.40 = 201.10 m

Fore sight to CP_2 = 1.70 m

∴ RL of CP_2 = 201.10 – 1.70 = 199.40 m

Back sight to CP_2 from L_3 = 1.30 m

∴ Plane of collimation at L_3 = 199.40 + 1.30 = 200.70 m

Fore sight to B = 1.85 m

∴ **RL of B** = 200.70 – 1.85 = **198.85 m** **Ans.**

If there are intermediate sight to the points E_1 and E_2, the RL of those points may be obtained by subtracting readings for E_1 and E_2 from the corresponding plane of collimations.

Booking and Reducing the Levels

The booking of readings and reducing the levels can be carried out systematically in the tabular form. There are two such methods:

(*i*) Plane of collimation method (*ii*) Rise and fall method.

For the above problem, with intermediate sights to E_1 = 0.80 m and E_2 = 0.70 m is illustrated below by the both methods.

Table 5.1 Booking and reducing levels by plane of collimation method

Station	*BS*	*Reading IS*	*FS*	*Plane of collimation*	*RL*	*Remarks*
A	1.35			201.35	200.00	Bench mark
E_1		0.80			200.55	Plinth of building
CP_1	1.40		1.65	201.10	199.70	CP_1
E_2		0.70			200.40	Plinth of building
CP_2	1.30		1.70	200.70	199.40	CP_2
B			1.85		198.85	B
Check Σ BS = 4.05 Σ BS – Σ FS = – 1.15 (Fall)			Σ FS = 5.20		Diff in RL of A and B = 198.85 – 200.00 = –1.15	

In this method note the following:

1. Plane of collimation for first setting

   = RL of BM + BS
2. Subtract IS from plane of collimation to get RL of intermediate station and subtract FS from plane of collimation to get RL of change point.
3. Add back sight to RL of change point to get new plane of collimation.
4. Check: Σ BS – Σ FS = RL of Last point – RL of first point.

If it is – ve, it is fall and if +ve it is rise.

Table 5.2 Booking and reducing level by rise and fall method

Station	*BS*	*IS*	*FS*	*Rise*	*Fall*	*RL*	*Remarks*
A	1.35					200.00	Bench mark
		0.80		0.55		200.55	E_1
	1.40		1.65		0.85	199.70	CP_1
		0.70		0.70		200.40	E_2
	1.30		1.70		1.00	199.40	CP_2
			1.85		0.55	198.85	B
	Σ BS = 4.05		Σ FS = 5.20	Σ Rise = 1.25	Σ Fall = 2.40		
Check: Σ BS – Σ FS = –1.15			Σ Rise – Σ Fall = –1.15			RL of last point – RL of first point = –1.15	

Note the following:

1. From A to E_1, difference = 1.35 – 0.80 = 0.55, rise
2. From E_1 to CP_1, difference = 0.80 – 1.65 = – 0.85, fall
3. From CP_1 to E_2, difference = 1.40 – 0.70 = 0.70, rise
4. From E_2 to CP_2, difference = 0.70 – 1.70 = –1.00, fall
5. From CP_2 to B, difference = 1.30 – 1.85 = – 0.55, fall.

5.8.3 Fly Levelling

If the work site is away from the bench mark, surveyor starts the work with a back sight on the bench mark by setting instrument at a convenient point. Then he proceeds towards the site by taking fore sights and back sights on a number of change points till he establishes a temporary bench mark in the site. Rest of the levelling work is carried out in the site. At the end of the work again levelling is carried out by taking a set of convenient change points till the bench work is reached. This type of levelling in which only back sight and fore sights are taken, is called fly levelling, the purpose being to connect a bench mark with a temporary bench mark or vice versa. Thus the difference between fly levelling and differential levelling is only in the purpose of levelling.

5.8.4 Profile Levelling

This type of levelling is known as longitudinal sectioning. In high way, railway, canal or sewage line projects profile of the ground along selected routes are required. In such cases, along the route, at regular interval readings are taken and RL of various points are found. Then the section of the route is drawn to get the profile. Figure 5.8 (*a*) shows the plan view of the scheme of levelling and Fig. 5.8 (*b*) shows the profile of the route. For drawing profile of the route, vertical scale is usually larger compared to scale for horizontal distances. It gives clear picture of the profile of the route.

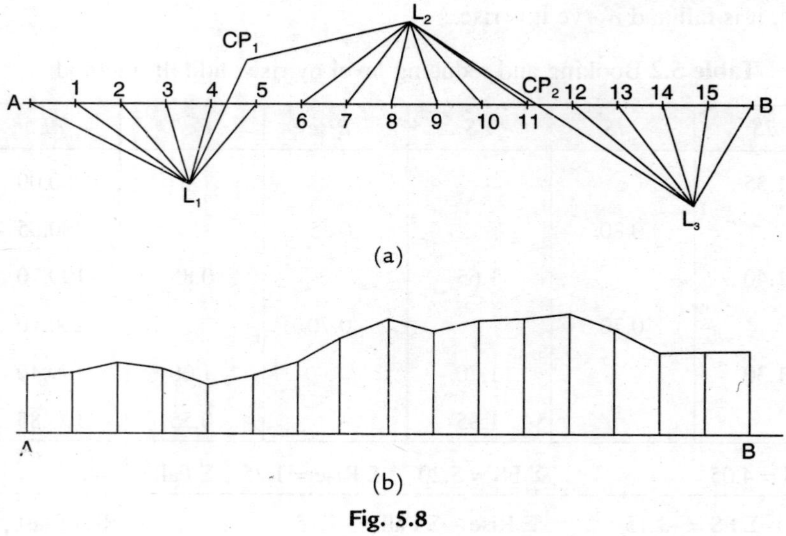

Fig. 5.8

The typical page of field book for this work will be having an additional column to note distances as shown in Table 5.3.

Table 5.3 Page of level book for profile levelling

Station	*Distance*	*BS*	*IS*	*FS*	*Plane of collimation*	*RL*	*Remark*

5.8.5 Cross-Sectioning

In many engineering projects, not only longitudinal profile but also the profile of cross-sections at regular intervals are required. These profiles help in calculating the earth works involved in the projects. Figure 5.9 shows the scheme of such work in which longitudinal profile is found by taking readings at 20 m interval along chain lines AB, BC and readings are taken at an interval of 3 m on either side. The distances on the cross-sections are treated as left or right of the lines as they are found while facing the forward station of survey. The cross-sectional length depends upon the nature of the project.

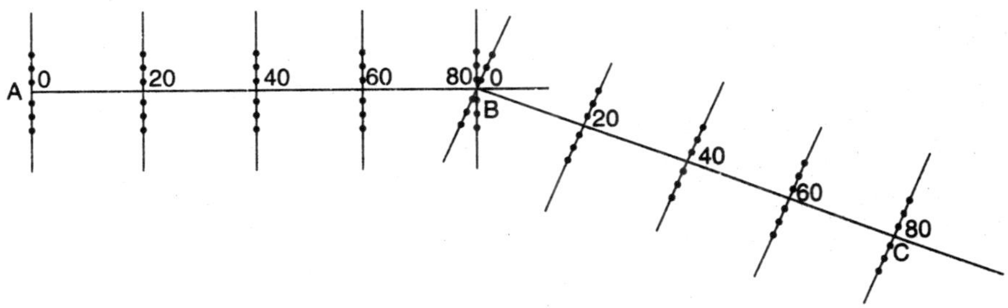

Fig. 5.9

Table 5.4 shows a page of level book required for this type of levelling.

Table 5.4 A typical page book for cross-section levelling

Station	*Distance in m*			*Readings*			*Plane of collimation*	*RL*	*Remarks*
	L	*C*	*R*	*BS*	*IS*	*FS*			
BM									
		0							
L_1	3								
L_2	6								
L_3	9								
R_1			3						
R_2			6						

R_3			9						
		20							
L_1	3								
L_2	6								
L_3	9								
R_1			3						
R_2			6						
R_3			9						
									Checked.

5.8.6 Reciprocal Levelling

In levelling, it is better to keep distance of back sight and fore sight equal. By doing so the following errors are eliminated.

(*i*) Error due to non-parallelism of line of collimation and axis of bubble tube.

(*ii*) Errors due to curvature and refraction. But in levelling across obstacles like river and ravine, it is not possible to maintain equal distances for fore sight and back sight. In such situations reciprocal levelling as described below is used:

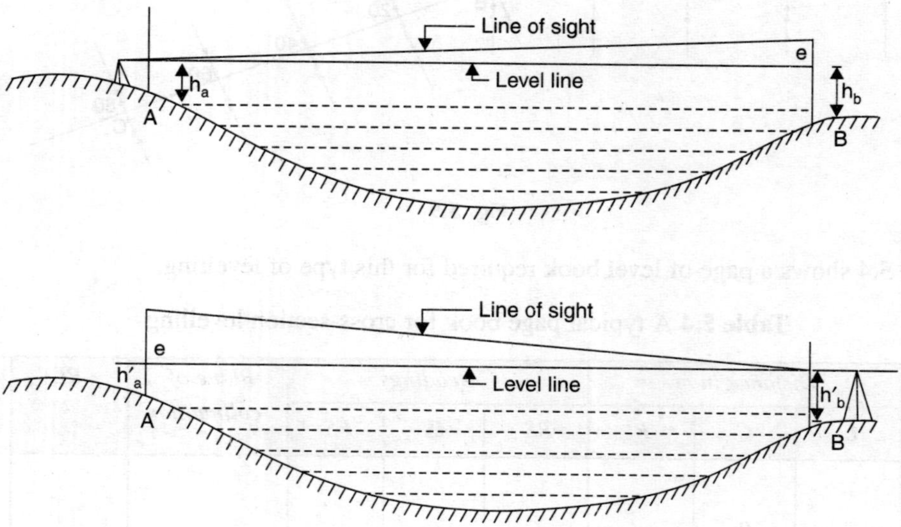

Fig. 5.10

(*i*) Referring to Fig. 5.10 (*a*).

Since A is very close, error in reading at A is negligible. Hence h_a is correct reading.

Let error in h_b be '*e*',

Then correct reading at B = $h_b - e$

∴ Difference in elevations = H = $h_a - (h_b - e)$ ... (*i*)

(*ii*) Referring to Fig. 5.10 (*b*), since B is very close to instrument, h_b', can be taken as correct reading.

Correct reading at A = $h_a' - e$

Difference in elevations H = $(h_a' - e) - h_b'$ ... (*ii*)

From equations (*i*) and (*ii*) we get,

$$2\,H = h_a - (h_b - e) + (h_a' - e) - h_b'$$
$$= (h_a + h_a') - (h_b + h_b')$$

$$\therefore \quad H = \frac{(h_a + h_a') - (h_b + h_b')}{2} \qquad ... (5.1)$$

Thus, the true difference in the elevations of the two points is equal to the mean of the two apparent differences in the elevations.

■ **Example 5.1:** *The following staff readings were observed successively with a level. The instrument has been shifted after the second and fifth reading: 0.675, 1.230, 0.750, 2.565, 2.225, 1.935, 1.835, 3.220. The first reading was with staff held on bench mark of RL 100.000 m. Enter the readings in a page of level book and calculate the RL of all points. Apply arithmetic checks. Use plane of collimation method.*

Solution: It is carried out as shown in Table 5.5.

Table 5.5

Station	*BS*	*IS*	*FS*	*Plane of collimation*	*RL*	*Remarks*
	0.675			100.675	100.00	BM of RL = 100.00
	0.750		1.230	100.195	99.445	CP_1
		2.565			97.630	
	1.935		2.225	99.905	97 970	CP_2
		1.835			98 070	
			3.220		96 685	Last Point
	Σ BS = 3.360		Σ FS = 6.675		Last RL – first RL	
	Σ BS –Σ FS = – 3.315 (Fall)				= – 3.315 (Fall)	Checked

■ **Example 5.2:** *Reduce the levels in example 5.1 by rise and fall method.*

Solution: It is carried out as shown in Table 5.6 below:

Table 5.6

Station	*BS*	*IS*	*FS*	*Rise*	*Fall*	*RL*	*Remarks*
	0.675					100.00	BM
	0.750		1.230		0.555	99.445	CP_1
		2.565			1.815	97.630	
	1.935		2.225	0.340		97.970	CP_2
		1.835		0.100		98.070	
			3.220		1.385	96.685	Last point
Σ	3.360		6.675	0.440	3.755		
∴	Σ BS – Σ FS = – 3.315		Σ Rise – Σ Fall = – 3.315		RL of last – RL of first point = – 3.315 Checked.		

■ **Example 5.3:** *The following consecutive readings were taken with a level on a sloping ground at a common interval of 20 m.*

0.600, 1.235, 1.860, 2.575, 0.235, 0.915, 1.935 and 2.870.

The reduced level of the first point was 192.125 m. Rule out a page of level book and enter the above readings. Calculate the reduced levels of points. Apply the check.

Solution: Table 5.7 shows entries in typical page of level book and method of reducing the levels.

Table 5.7

Station	*Distance*			*Readings*			*Plane of collimation*	*RL*	*Remarks*
	L	*C*	*R*	*BS*	*IS*	*FS*			
A		0.0		0.600			192.725	192.125	First point
		20.0			1.235			191.149	
		40.0			1.860			190.865	
		60.0			2.575			190.150	
		80.0			0.235			192.490	
		100.0			0.915			191.810	
		120.0				2.870		189.855	Last point
			Σ BS = 0.600 Σ BS – Σ IS = – 2.270		Σ FS = 2.870		RL of last point – RL of first point = – 2.27		Checked

■ **Example 5.4:** *Reciprocal levelling was conducted with a dumpy level and the following readings were recorded.*

Instrument near	*Staff reading at*	
	A	*B*
A	*1.245*	*1.575*
B	*1.050*	*0.700*

Compute the RL of B, if RL of A = 218.250 m.

Solution: $h_a = 1.245$ m $\quad h_b = 1.575$ m

$h_a' = 1.050$ m $\quad h_b' = 0.700$ m

∴ Difference in levels of A and B is

$$= \frac{(h_a - h_b) + (h_a' - h_b')}{2}$$

$$= \frac{(1.245 - 1.575) + (1.050 - 0.700)}{2} = 0.01$$

Now plane of collimation at A = 218.250 + 1.245

= 219.495 m

∴ **Elevation of B** = 219.495 – 0.01

= **219.485 m Ans.**

5.9 CONTOURS

A contour line is a imaginary line which connects points of equal elevation. Such lines are drawn on the plan of an area after establishing reduced levels of several points in the area. The contour lines in an area are drawn keeping difference in elevation of between two consecutive lines constant. For example Fig. 5.11 shows contours in an area with contour interval of 1 m. On contour lines the level of lines is also written.

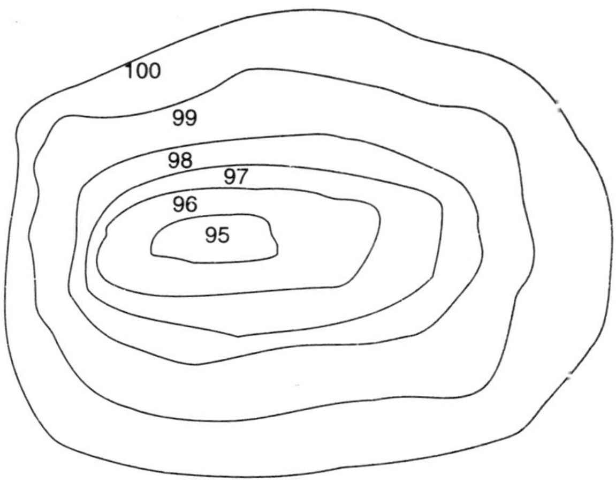

Fig. 5.11 Contours

5.9.1 Characteristics of Contours

The contours have the following characteristics:

1. Contour lines must close, not necessarily in the limits of the plan.
2. Widely spaced contour indicates flat surface.

3. Closely spaced contour indicates steep ground.
4. Equally spaced contour indicates uniform slope.
5. Irregular contours indicate uneven surface.
6. Approximately concentric closed contours with decreasing values towards centre (Fig. 5.11) indicate a pond.
7. Approximately concentric closed contours with increasing values towards centre indicate hills.
8. Contour lines with U-shape with convexity towards lower ground indicate ridge (Fig. 5.12).

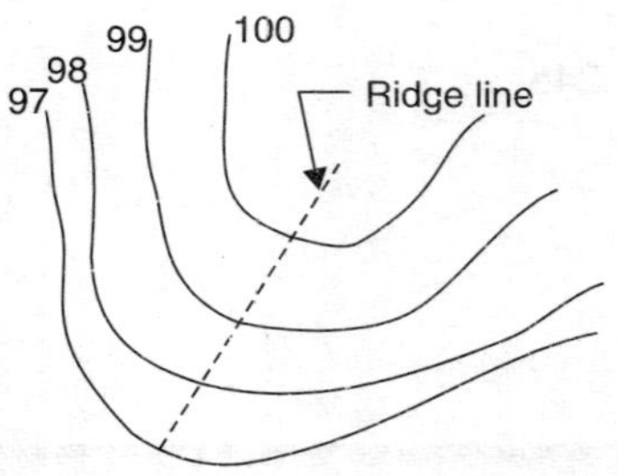

Fig. 5.12

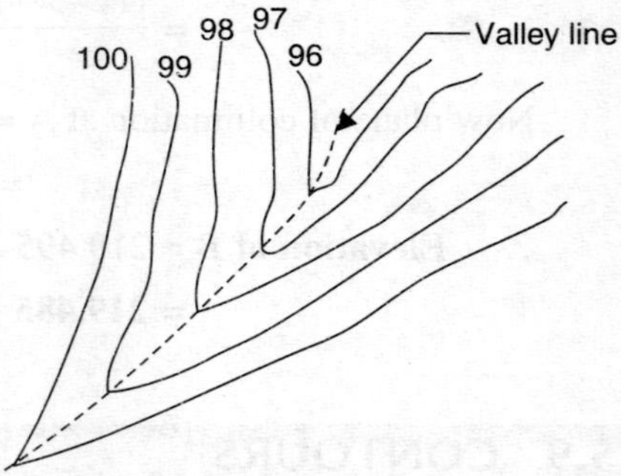

Fig. 5.13

9. Contour lines with V-shaped with convexity towards higher ground indicate valley (Fig. 5.13).
10. Contour lines generally do not meet or intersect each other.
11. If contour lines are meeting in some portion, it shows existence of a vertical cliff (Fig. 5.14).

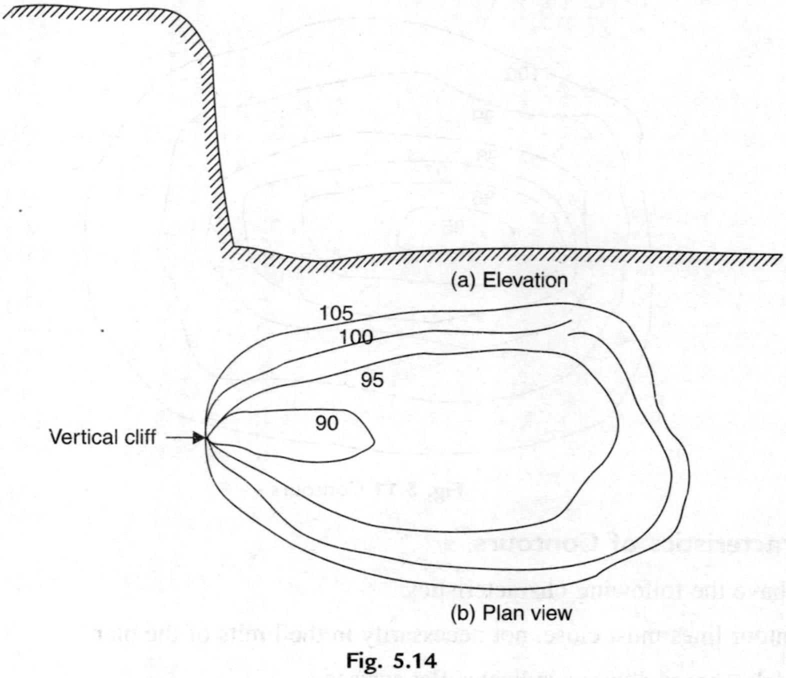

Fig. 5.14

12. If contour lines cross each other, it shows existance of overhanging cliffs or a cave (Fig. 5.15).

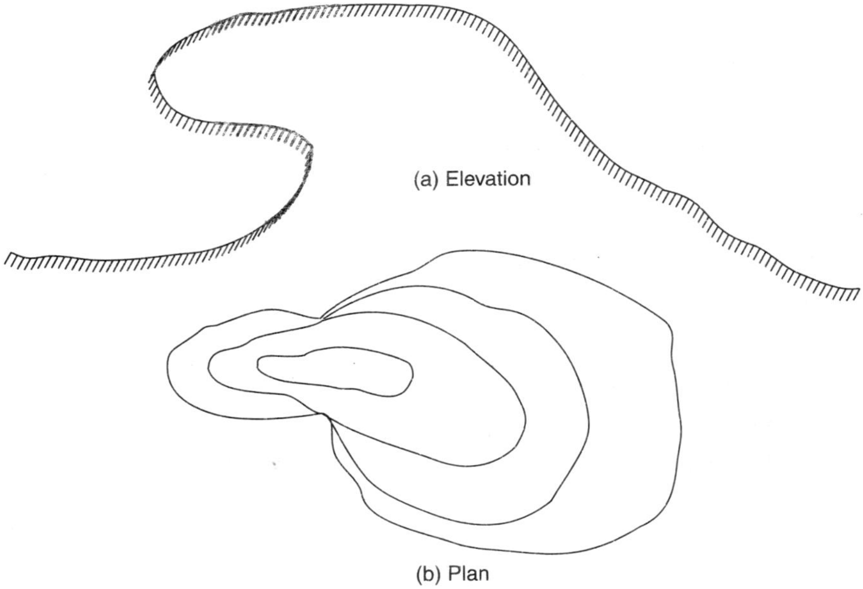

Fig. 5.15

5.9.2 Uses of Contour Maps

Contour maps are extremely useful for various engineering works:

1. A civil engineer studies the contours and finds out the nature of the ground to identify. **suitable site for the project** works to be taken up.
2. By **drawing the section** in the plan, it is possible to find out profile of the ground along that line. It helps in finding out depth of cutting and filling, if formation level of road/railway is decided.
3. **Intervisibility of any two points** can be found by drawing profile of the ground along that line.
4. **The routes** of the railway, road, canal or sewer lines can be decided so as to minimize and balance earthworks.
5. **Catchment area** and hence quantity of water flow at any point of nalla or river can be found. This study is very important in locating bunds, dams and also to find out flood levels.
6. From the contours, it is possible to determine the **capacity of a reservoir.**

QUESTIONS

1. Define the following terms used in levelling:

(*i*) Bench mark (*ii*) Level line

(*iii*) Change point (*iv*) Mean sea level.

2. Differentiate between:

   (*i*) GTS bench mark and Temporary bench mark

   (*ii*) Fore sight and back sight

   (*iii*) Plane of collimation and RL of a point

   (*iv*) Fly levelling and profile levelling.

3. What are temporary adjustment of dumpy level ? Explain how they are performed.
4. Write short notes on Barometric levelling and Hypsometry.
5. Explain differential levelling.
6. Write short note on reciprocal levelling.
7. Explain the terms: contour lines and contour interval.
8. List the characteristics of contours.
9. Sketch the contours for the following:

   (*i*) Valley (*ii*) Ridge

   (*iii*) Vertical cliff (*iv*) Overhanging cliff.

10. List the uses of contour maps.
11. The following staff readings were taken with a level, the instrument having been moved after third and sixth reading:

    2.200 1.620 0.980 2.250 2.840 1.280 0.600 1.960 1.450

    The RL of first point is 100.00 m. Rule out a page of level book, enter the above readings. Calculate the RL of all points. Apply the checks. **[Ans: RL of last point 101.34.**

    Note. First, fourth and seventh readings are back sights. Third, sixth and last readings are fore sights]

6

Areas and Volumes

The land is always bought and sold on the basis of cost per unit area. For road and railways land is to be acquired on the basis of area. In the design of bridges and bunds catchment area of river and nalla are required. Thus finding areas is the essential part of surveying. It may be noted that area to be found is the projected area. Units used for finding areas are square metres, hectare and square kilometre. Relation among them are

$$\text{Hectare} = 100 \text{ m} \times 100 \text{ m} = 1 \times 10^4 \text{ m}^2$$

$$\text{Square kilometer} = 1000 \text{ m} \times 1000 \text{ m} = 1 \times 10^6 \text{ m}^2$$

$$= 100 \text{ hectare}$$

Similarly volume of earth work involved in projects like road, rail and canal are to be found by surveying. Capacity of reservoir also need volume calculations. In this chapter calculation of areas and volumes based on surveying are explained.

6.1 COMPUTATION OF AREAS FROM FIELD NOTES

If the area is bound by straight edges, it can be subdivided in a set of convenient figures and area calculated. But in most of the cases the boundary may have irregular shape. In such cases major area is subdivided into regular shape and area is found. The smaller area near the boundary is found from taking offset from a survey line [Ref. Fig. 6.1].

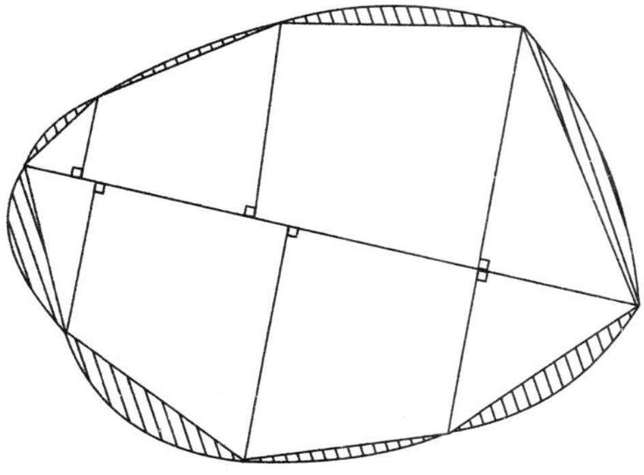

Fig. 6.1

6.1.1 Computation of Areas of Regular Figures

The following expressions for calculating areas may be noted:

(*a*) **Triangle:**

(*i*) If base width is b and height is 'h',

$$A = \frac{1}{2} bh \qquad \text{... (6.1)}$$

(*ii*) If a, b and c are the sides of a triangle,

$$A = \sqrt{s(s-a)(s-b)(s-c)}$$

where $$s = \frac{a+b+c}{2} \qquad \text{... (6.2)}$$

(*b*) **Rectangle:** If b and 'd' are the dimension of a rectangle,

$$A = bd \qquad \text{.... (6.3)}$$

(*c*) **Trapezium:**

$A = d\,\dfrac{h_1 + h_2}{2}$, where d is the distance between two parallel sides and h_1 and h_2 lengths of parallel sides. .... (6.4)

6.1.2 Areas of Irregular Shapes

For this purpose from a survey line offsets are taken at regular intervals and area is calculated from any one of the following methods:

(*a*) Area by Trapezoidal rule

(*b*) Area by Simpson's rule.

(*a*) **Area by Trapezoidal Rule:**

If there are '$n + 1$' ordinates at n equal distances 'd', then total length of line is $L = nd$, Area of each segment is calculated treating it as a trapezium. Referring to Fig. 6.2,

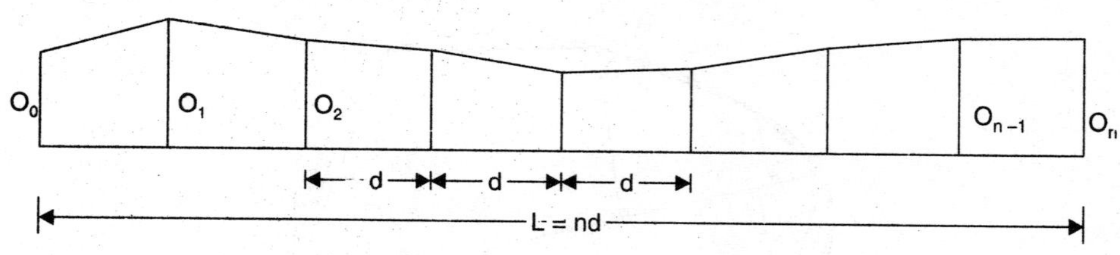

Fig. 6.2

$$\text{Area of first segment} = \frac{O_0 + O_1}{2}\, d$$

By adding all such segmental areas, we get total

$$A = \frac{O_0 + O_1}{2} d + \frac{O_1 + O_2}{2} d + \cdots\cdots + \frac{O_{n-1} + O_n}{2} d$$

$$= \left[\frac{O_0 + O_n}{2} + O_1 + O_2 + \cdots + O_{n-1}\right] d \qquad \ldots (6.5)$$

(*b*) **Area by Simpson's Rule**

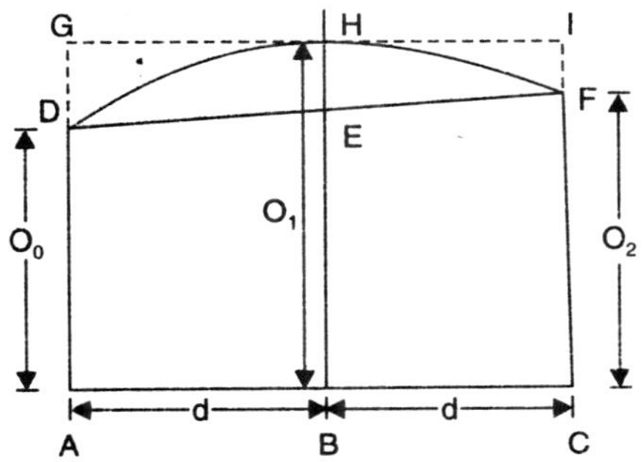

Fig. 6.3

In this method, the boundary line between two segment is assumed parabolic. Figure 6.3 shows the first two segments of Fig. 6.2, in which boundary between the ordinates is assumed parabolic.

∴ Area of the first two segments

= Area of trapezium ACFD + Area of parabola DEFH

$$= \frac{O_0 + O_2}{2} 2d + \frac{2}{3} \times 2d \times EH$$

$$= (O_0 + O_2)\, d + \frac{4}{3} d \left(O_1 - \frac{O_0 + O_2}{2}\right)$$

$$= \frac{d}{3}[3O_0 + 3O_2 + 4O_1 - 2O_0 - 2O_2]$$

$$= \frac{d}{3}[O_0 + 4O_1 + O_2]$$

Area of next two segments

$$= \frac{d}{3}[O_2 + 4O_3 + O_4]$$

Area of last two segments

$$= \frac{d}{3}[O_{n-2} + 4O_{n-1} + O_n]$$

∴ Total

$$A = \frac{d}{3}\left[\begin{matrix}(O_0 + O_n) + 4(O_1 + O_3 + \cdots + O_{n-1}) \\ + 2(O_2 + O_4 + \cdots + O_{n-2})\end{matrix}\right] \qquad \ldots. (6.6)$$

It is to be noted that the equation 6.6 is applicable if the number of segments (*n*) are even, in other words, if total number of ordinates's are odd.

If n is odd, then for $n - 1$ segments area is calculated by Simpson's rule and for the last segment trapezoidal rule is applied.

Trapezoidal rule gives better results if the boundary is not irregular to great extent. Simpson's rule should be used if the boundary is highly irregular. This rule gives slightly more value compared to trapezoidal rule, if the curve is concave towards the survey line and gives lesser value, if the boundary is convex towards survey line.

In both methods accuracy can be improved if the number of segments are increased.

■ **Example 6.1:** *Plot the following cross staff survey of a land and compute the area:*

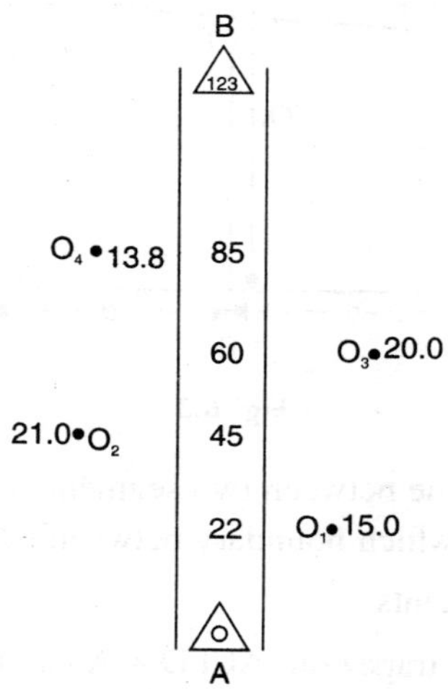

Solution: Referring to Fig. 6.4,

Total area = Area of Δ AO_1C_1 + Area of triangle AC_2O_2 + Area of trapezium $C_1O_1O_3C_3$ + Area of trapezium $C_2C_4O_4O_2$ + Area of Δ C_3O_3B + Area of Δ C_4O_4B

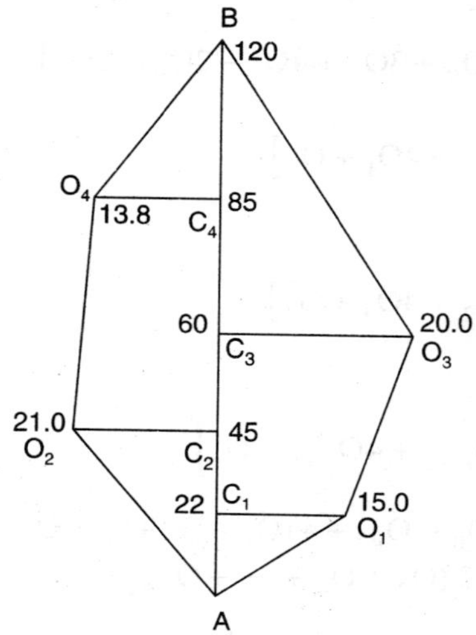

Fig. 6.4

$$= \frac{1}{2} \times 22 \times 15 + \frac{1}{2} \times 45 \times 21 + \frac{1}{2} (15 + 20)(60 - 22) + \frac{1}{2} (13.8 + 21.0)(85 - 45)$$

$$+ \frac{1}{2} \times 20 (120 - 60) + \frac{1}{2} \times 13.8 (120 - 8.5)$$

Area = 2840 m² **Ans.**

■ **Example 6.2:** *The perpendicular offsets taken at 10 m intervals from a survey line to an irregular boundary are 2.18 m, 3.2 m, 4.26 m, 6.2 m, 4.8 m, 7.20 m, 8.8 m, 8.2 m and 5.2 m. Determine the area enclosed between the boundary, survey line, the first and the last offsets by*

(i) Trapezoidal rule *(ii) Simpson's rule.*

Solution: $d = 10$ m, n = number of segments = 8 number of ordinates = 9.

Length of survey line = 8 × 10 = 80 m.

(*i*) **Area by trapezoidal rule**

$$A = \left(\frac{O_0 + O_8}{2} + O_1 + O_2 + \cdots + O_7\right) d$$

$$= \left[\frac{2.18 + 5.2}{2} + 3.2 + 4.26 + 6.2 + 4.8 + 7.2 + 8.8 + 8.2\right] 10$$

Area = 463.5 m² **Ans.**

(*ii*) **Area by Simpson's method**

$$= \frac{d}{3} [(O_0 + O_8) + 4(O_1 + O_3 + O_5 + O_7) + 2(O_2 + O_4 + O_6)]$$

$$= \frac{10}{3} [2.18 + 5.2 + 4(3.2 + 6.2 + 7.2 + 8.2) + 2(4.26 + 4.8 + 8.8)]$$

= 474.333 m² **Ans.**

■ **Example 6.3:** *The following offsets were taken to a curved boundary from a survey line:*

0, 2.46, 3.78, 3.26, 4.40, 3.28, 4.24 and 5.20 m.

Compute the area between curved boundary, survey line and end offsets, if the offsets were at a regular interval of 10 m, using Simpson's rule and trapezoidal rule. Compare the results.

Solution: Number of offsets = 8

Number of intervals = 7

$O_0 = 0.0$, $O_1 = 2.40$, $O_2 = 3.78$, $O_3 = 3.26$, $O_4 = 4.40$, $O_5 = 3.28$,

$O_6 = 4.27$, $O_7 = 5.20$

$d = 10.0$ m

(*i*) **From trapezoidal rule**

$$\mathbf{A} = \left[\frac{O_0 + O_7}{2} + O_1 + O_2 + O_3 + O_4 + O_5 + O_6\right] d$$

$$= \left[\frac{0.0 + 5.20}{2} + 2.46 + 3.78 + 3.26 + 4.40 + 3.28 + 4.27\right]10$$

$= \mathbf{240.5\ m^2}$ **Ans.**

(*ii*) **Simpson's Rule:**

Number of intervals are odd (7). Hence for first six intervals Simpson's rule can be applied and for the last interval, trapezoidal rule will be applied.

$$A = \frac{d}{3}\,[O_0 + O_6 + 4\,(O_1 + O_3 + O_5) + 2\,(O_2 + O_4)] + \frac{d}{2}\,(O_6 + O_7)$$

$$= \frac{10}{3}\,[0 + 4.27 + 4(2.46 + 3.26 + 3.28) + 2(3.78 + 4.40)] + \frac{10}{2}\,(4.27 + 5.20)$$

$= \mathbf{236.12\ m^2}$ **Ans.**

$\therefore$ Simpson's rule gave lesser area, the magnitude being 240.5 – 236.12 = **3.38 m²**. **Ans.**

6.2 COMPUTING AREAS FROM MAPS

If map of an area is available its area can be found by the following methods:

(*i*) Approximate methods (*ii*) Using planimeter.

6.2.1 Approximate Methods

The following three approximate methods are available for calculating area from the map:

(*a*) Give and take method (*b*) Subdivisions into squares

(*c*) Subdivisions into trapezoids.

(*a*) **Give and take method:** In this method irregular boundary is approximated with straight lines such that area taken in is equal to the area given out. Accuracy depends upon the judging capacity of the engineer. Then the area with straight edges is divided into a set of simple figures, like triangles and trapezoids and the area of map is found using standard expressions. Figure 6.5 shows such a scheme.

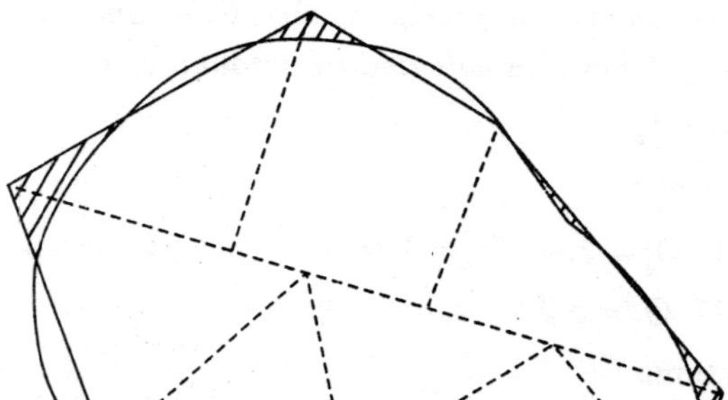

Fig. 6.5

(*b*) **Subdivisions into squares:** Similar to a graph sheet, squares are marked on a transparent tracing sheet, each square representing a known area. Full squares are counted. Fractional squares are counted by give and take approximation. Then the number of squares multiplied by area of each square gives the area of the map Fig. 6.6 shows such a scheme. Finer the mesh better is the accuracy.

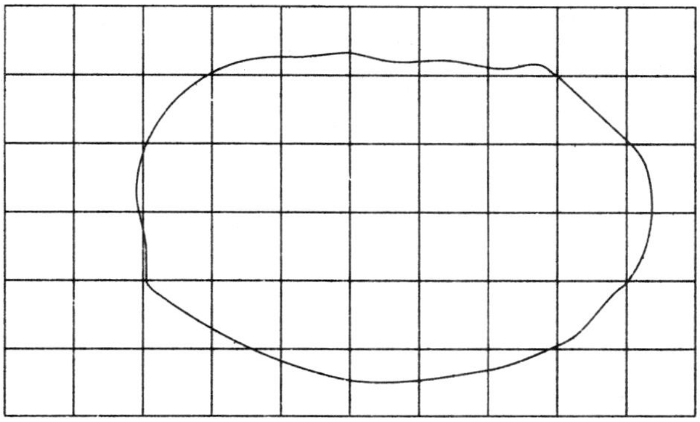

Fig. 6.6

(*c*) **Subdivisions into rectangles:** In this method, a set of parallel lines are drawn at equal spacing on a transparent paper. Then that sheet is placed over the map and slightly rotated till two parallel lines touch the edges of the tap. Then equalising perpendiculars are drawn between the consecutive parallel line. Thus given area is converted to equivalent set of rectangles and then area is calculated (Ref. Fig. 6.7).

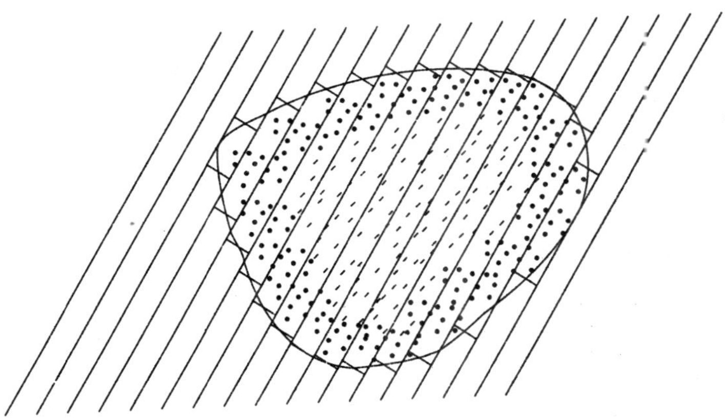

Fig. 6.7

6.2.2 Computing Area Using Planimeter

Planimeter is a mechanical instrument used for measuring area of plan. The commonly used planimeter is known as Amsler planimeter (Fig. 6.8). Its construction and uses are explained in this article.

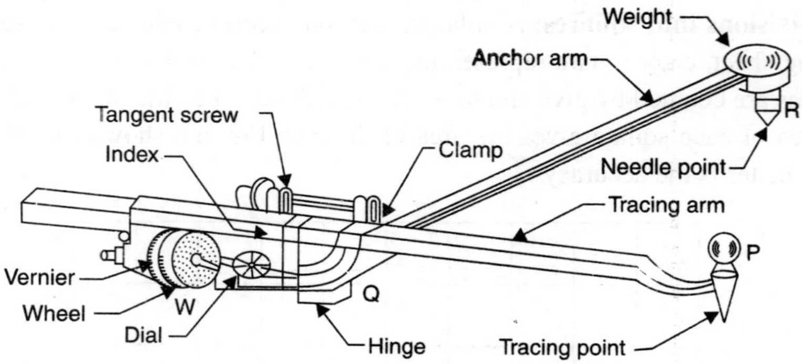

Fig. 6.8 Planimeter

The essential parts of a planimeter are:

1. Anchor: It is a heavy block with a fine anchor pin at its base. It is used to anchor the instrument at a desired point on the plan.
2. Anchor arm: It is a bar with one end attached to anchor block and the other connected to an integrating unit. Its arm length is generally fixed but some planimeters are provided with variable arms length also.
3. Tracing arm: It is a bar carrying a tracer point at one end connected to the integrating unit at the other end. The anchor arm and tracer arms are connected by a hinge. The length of this arm can be varied by means of fixed screw and slow motion screw.
4. Tracing point: This is a needle point connected to the end of tracer arm, which is to be moved over the out line of the area to be measured.
5. Integrating unit: It consists of a hard steel roller and a disc. The axis of roller coincides with the axis of tracer arm hence it rolls only at right angles to the tracer arm. The roller carries a concentric dream which has 100 divisions and is provided with a vernier to read tenth of roller division. A suitable gear system moves a pointer on disc by one division for every one revolution of the roller. Since the disc is provided with 10 such equal divisions, the reading on the integrating unit has four digits:

   (*i*) Unit read on the disc

   (*ii*) Tenth and hundredth of a unit read on the roller

   (*iii*) Thousandth read on the vernier.

Thus if reading on disc is 2, reading on roller is 42 and vernier reads 6, then the total reading

$$F = 2.426$$

Method of Using Planimeter

To find the area of a plan, anchor point may be placed either outside the plan or inside the plan. It is placed outside the plane, if the plane area is small. Then on the boundary of the plan a point is marked and tracer is set on it. The planimeter reading is taken. After this tracer is carefully moved over the outline of the plan in clockwise direction till the first point is reached. Then the reading is noted. Now the area of the plan may be found as

$$\text{Area} = M\,(F - I + 10\,N + C) \quad \text{.... (6.7)}$$

where M = A multiplying constant

F = Final reading

I = Initial reading.

N = The number of completed revolutions of disc. Plus sign to be used if the zero mark of the dial passes index mark in clockwise direction and minus sign if it passes in anticlockwise direction.

C = Constant of the instrument, which when multiplied with M, gives the area of zero circle.

The constant C is added only when the anchor point is inside the area.

Multiplying constant M is equal to the area of the plan (map) per revolution of the roller *i.e.,* area corresponding to one division of disc.

Multiplying constant M and C are normally written on the planimeter. The user can verify these values by

(*i*) Measuring a known area (like that of a rectangle) keeping anchor point outside the area

(*ii*) Again measuring a known area by keeping anchor point inside a known area.

The method is explained with example.

The proof of equation 6.7 is considered as beyond the scope of this book. Interested readers can see the book on surveying and levelling.

■ **Example 6.4:** *To determine the constants of a planimeter a 20 cm × 20 cm area was measured with anchor point outside the plan area. The zero mark of disc crossed the index in clockwise direction once. The observed readings are*

Initial reading = 7.422

Final reading = 1.422

Determine the multiplying constant M.

Solution: Area = M (F – I + 10 N)

Now Area = 20 × 20 = 400 cm^2

F = 1.422, I = 7.422

∴ 400 = M (1.422 – 7.422 + 10 × 1)

∴ **M = 100 cm^2** **Ans.**

■ **Example 6.5:** *A planimeter was used to measure the area of a map once keeping anchor point outside the figure and second time keeping it inside the figure. The observations are as follows:*

(*i*) *When the anchor point was outside the figure:*

Initial reading = 1.486

Final reading = 7.058

The zero of the dial did not pass the index at all.

(*ii*) *When the anchor point was inside the map:*

Initial reading = 3.486

Final reading = 8.844

Zero mark of the dial passed the fixed index mark twice in anticlockwise direction.

Find the (i) area of the map (ii) area of the zero circle.

Take multiplier constant $M = 100\ cm^2$.

Solution: (*i*) When the anchor point was outside the plan

$$I = 1.486 \quad F = 7.058 \quad N = 0 \quad M = 100\ cm^2.$$

$$\therefore \quad A = M\,(F - I + 10\,N)$$

$$= 100\,(7.058 - 1.486 + 0)$$

$$\therefore \quad \mathbf{A = 557.2\ cm^2} \qquad \textbf{Ans.}$$

(*ii*) When the anchor point was inside the plan

$$I = 3.486 \quad F = 8.844 \quad N = -2 \quad M = 100\ cm^2.$$

$$A = M\,(F - I + 10\,N + C)$$

$$557.2 = 100\,(8.844 - 3.486 - 10 \times 2 + C)$$

$$\therefore \quad \mathbf{C = 20.214} \qquad \textbf{Ans.}$$

6.3 COMPUTATION OF VOLUMES

The following three methods are available for computation of volumes:

(*i*) From cross-sections (*ii*) From spot levels and

(*iii*) From contours.

First method is useful for computing earth work involved in road/rail/canal/sewage works. Second method is useful for finding earth work in foundations of large building and the last method is useful for finding capacity of reservoirs.

6.3.1 Computation of Volume from Cross-sections

To compute earth work, profile levelling is carried out along the centre line of the alignment of the project and cross-sectional levels are taken at regular intervals. Then the volume of earth work can be found, if the cross-sections are determined.

First the calculation of cross-sectional area is discussed.

(*a*) **If section is level (Fig. 6.9)**

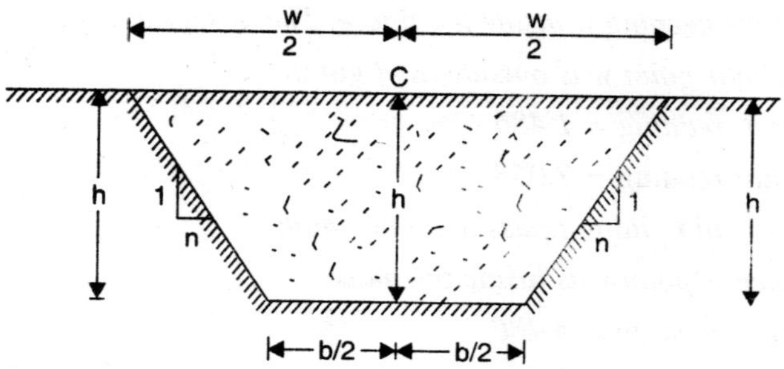

Fig. 6.9

Let 'h' be the depth at the centre line of the alignment and 1 : n be the side slopes. Then

$$w = b + 2nh$$

$$\therefore \qquad A = \frac{1}{2}(w + b)\,h$$

$$= \frac{1}{2}(b + 2nh + b)\,h$$

$$= (b + nh)\,h \qquad \text{... (6.8)}$$

(b) If it is a multilevel section [Fig. 6.10]

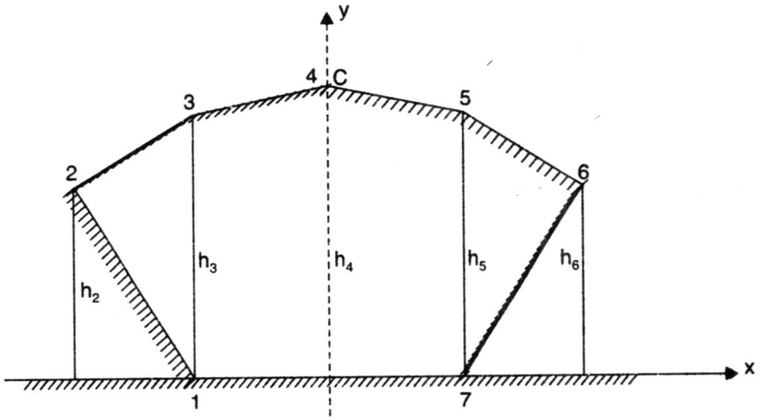

Fig. 6.10

Let the coordinates of points be $(x_1, y_1), (x_2, y_2), ..., (x_n, y_n)$, then arrange the coordinates in the following order

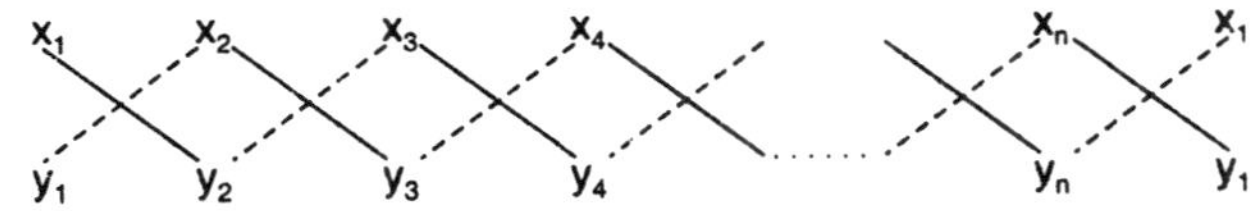

Then area of the figure

$= \frac{1}{2}$ [Σ Product of pair of coordinates connected by continuous lines – Σ Product of coordinates connected by dotted lines] ... (6.9)

The above formula can be easily proved by taking a simple example of a quadrilateral [Ref. Fig. 6.11]. Let the coordinates of A, B, C and D be (x_1, y_1), (x_2, y_2), (x_3, y_3) and (x_4, y_4). Then area of ABCD

= Area of a AB b + Area of b BC c + Area of c CD d – Area of a ADd.

$$= \frac{1}{2}(x_1 + x_2)(y_2 - y_1) + \frac{1}{2}(x_2 + x_3)(y_3 - y_2) + \frac{1}{2}(x_3 + x_4)(y_4 - y_3) - \frac{1}{2}(x_1 + x_4)(y_4 - y_1)$$

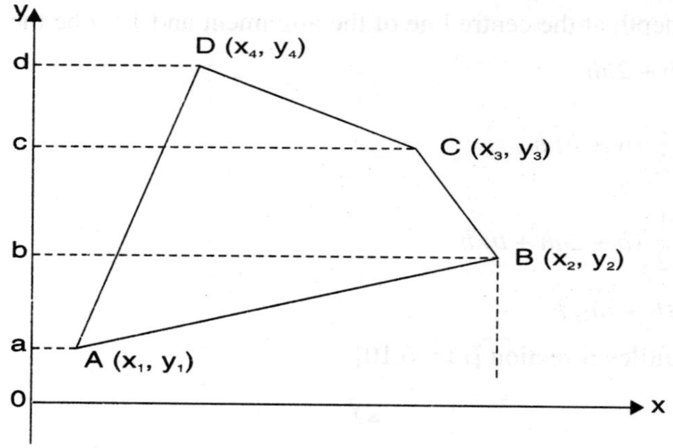

Fig. 6.11

$$= \frac{1}{2}\,[x_1y_2 - x_1y_1 + x_2y_2 - x_2y_1 + x_2y_3 - x_2y_2 + x_3y_3 - x_3y_2 + x_3y_4 - x_3y_3 + x_4y_4 - x_4y_3 - x_1y_4 + x_1y_1 - x_4y_4 + x_4y_1]$$

$$= (x_1y_2 + x_2y_3 + x_3y_4 + x_4y_1) - (x_2y_1 + x_3y_2 + x_4y_1 + x_1y_4)$$

[Note terms with same subscript appear in pairs and cancell each other].

Hence equation 6.9 is proved.

Calculation of Volumes

Once cross-sectional areas at various sections are known volume can be found from trapezoidal or prismoidal rule as given below:

Trapezoidal Rule

$$V = d\left[\frac{A_0 + A_n}{2} + A_1 + A_2 + \cdots + A_{n-1}\right] \quad \text{... (6.10)}$$

where 'n' are number of segments at interval of 'd', Area at L = nd, being A_n.

Prismoidal Rule

$$V = \frac{d}{3}\,[(A_0 + A_n) + 4\,(A_1 + A_3 + \cdots + A_{n-1}) + 2\,(A_2 + A_4 + \cdots + A_{n-2})] \quad \text{... (6.11)}$$

where n is number of even segments.

If number of segments are odd, (n is odd), for $n - 1$ segments prismoidal rule may be applied and for the last one trapezoidal rule is applied. Or else for the last segment area at middle of last segment found and prismoidal formula applied for A_{n-1}, A_m and A_n.

■ **Example 6.6:** *A railway embankment of formation width 12 m is to be built with side slopes of 1 vertical to 1.5 horizontal. The ground is horizontal in the direction transverse to the centre line. Length of embankment is 200 m. The centre height of embankment at 25 m interval are as given below :*

1.6, 2.4, 3.4, 3.8, 4.2, 3.6, 2.8, 2.2, 1.2 m. Calculate the volume of earth filling.

Solution: Since the section is level,

$$A = (b + nh)\,h \text{ where } n = 1.5$$

The area at different sections are:

$$A_0 = (12 + 1.5 \times 1.6)\ 1.6 = 23.04 \text{ m}^2$$
$$A_1 = (12 + 1.5 \times 2.4)\ 2.4 = 37.44 \text{ m}^2$$
$$A_2 = (12 + 1.5 \times 3.4) \times 3.4 = 58.14 \text{ m}^2$$
$$A_3 = (12 + 1.5 \times 3.8) \times 3.8 = 67.26 \text{ m}^2$$
$$A_4 = (12 + 1.5 \times 4.2) \times 4.2 = 76.86 \text{ m}^2$$
$$A_5 = (12 + 1.5 \times 3.6) \times 3.6 = 62.64 \text{ m}^2$$
$$A_6 = (12 + 1.5 \times 2.8) \times 2.8 = 45.36 \text{ m}^2$$
$$A_7 = (12 + 1.5 \times 2.2) \times 2.2 = 33.66 \text{ m}^2$$
$$A_8 = (12 + 1.5 \times 1.2) \times 1.2 = 16.56 \text{ m}^2$$

Volume by Trapezoidal Formula

$$V = d\left[\frac{A_0 + A_8}{2} + A_1 + A_2 + A_3 + A_4 + A_5 + A_6 + A_7\right]$$

$$= 25\left[\frac{23.04 + 16.56}{2} + 37.44 + 58.14 + 67.26 + 76.86 + 62.64 + 45.36 + 33.66\right]$$

$$= \mathbf{10029\ m^3}$$ **Ans.**

Volume by Prismoidal Formula

Since the number of segments are even (8), prismoidal formula

$$V = \frac{d}{3}\left[A_0 + A_8 + 4\,(A_1 + A_3 + A_5 + A_7) + 2\,(A_2 + A_4 + A_6)\right]$$

$$= \frac{25}{3}\left[23.04 + 16.56 + 4\,(37.44 + 67.26 + 62.64 + 33.66) + 2(58.14 + 76.86 + 45.36)\right]$$

$$= \mathbf{10036\ m^3}$$ **Ans.**

■ **Example 6.7:** *The following notes refer to a three level work in cutting.*

Station			
1	$\frac{2.0}{-7.0}$	$\frac{2.4}{0}$	$\frac{4.8}{11.0}$
2	$\frac{3.0}{-9.0}$	$\frac{3.8}{0}$	$\frac{6.0}{14.0}$

The formation level is in cutting and is 12 m wide. The distance between two points is 40 m. Calculate the volume of cutting between the two stations by

(i) Trapezoidal formula *(ii) Prismoidal formula.*

Solution: The meaning of the notes for station 1 is at centre point *i.e.,* at $x = 0$, depth of cutting is 2.4 m, at the end points on left at $x = 7$ m, the depth of cutting is 2 m and at a point $x = 11.0$ m depth is 4.8 m as shown in Fig. 6.11. Similarly the meaning for station 2 is also indicated in the Fig. 6.12.

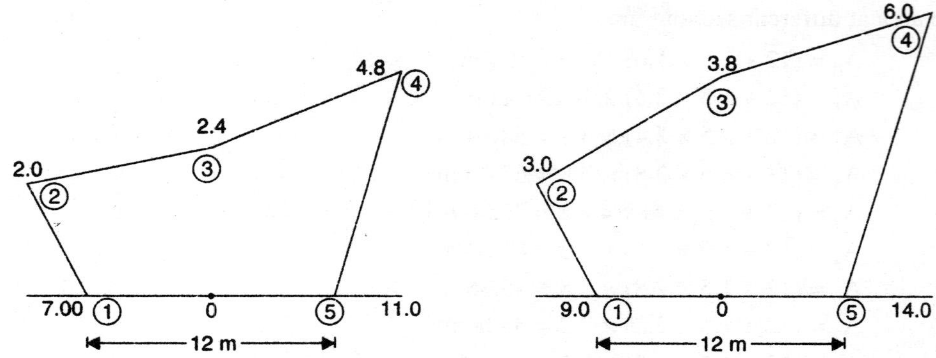

Fig. 6.12

Hence the coordinates of points (1) to (5) on first sections are:

Point →	1	2	3	4	5	1
x	−6.0	−7.0	0	11.0	6.0	−6.0
y	0	2.0	2.4	4.8	0	0

Note: Coordinates of first point are written again at the end is the coordinates method of area calculation.

$$A = \frac{1}{2}\ [\{-6 \times 2.0 + (-7.0 \times 2.4) + 0 \times 4.8 + 11.0 \times 0 + 6 \times 0\}$$

$$- \{(0.0)\ (-7.0) + 2 \times 0 + 2.4 \times 11.0 + 4.8 \times 6 + 0 \times (-6)\}]$$

$$= -42.0 \text{ m}^2$$

$= 42.0 \text{ m}^2$, since there is no meaning in –ve sign.

Similarly for section 2,

Point	1	2	3	4	5	1
x	− 6.0	− 9.0	0	14	6.0	− 6
y	0	3.0	3.8	6.0	0	0

$$\therefore \quad 2A = \{-6 \times 3.0 - 9.0 \times 3.8 + 0 \times 6 + 14 \times 0 + 6 \times 0\}$$

$$- \{0 \times (-9.0) + 3.0 \times 0 + 3.8 \times 14 + 6 \times 6.0 \times 0 \times (-6)\}$$

$$= -141.4$$

$$\therefore \quad A = 70.7 \text{ m}^2$$

From Trapezoidal Rule

$$V = \frac{1}{2} \times 40\ (42.0 + 70.7) = \mathbf{2254.0\ m^3} \qquad \textbf{Ans.}$$

Prismoidal Rule

To apply this we need cross-section at middle of the two sections. Assuming uniform slope, it may be found as the points with average coordinates. Hence at mid-section the coordinates of the points are:

Point	1	2	3	4	5	1
x	−6.0	−8.0	0	12.5	6.0	−6
y	0.0	2.5	3.1	5.4	0	0.0

$$\therefore \quad 2A_m = (-6 \times 2.5 - 8 \times 3.1 + 0 \times 5.4 + 12.5 \times 0 + 6 \times 0.0)$$
$$- \{0 \times (-8) + 2.5 \times 0 + 3.1 \times 12.5 + 5.4 \times 6.0 + 0 \times (-6)\}$$
$$= 110.95$$
$$A_m = 55.475 \text{ m}^2$$

Now $d = \dfrac{40}{2} = 20$ m, between two consecutive sections.

$$V = \frac{20}{3}(42.0 + 4 \times 55.475 + 70.7)$$

$$\mathbf{V = 2230.67\ m^3}$$ **Ans.**

6.3.2 Computation of Earth work From Spot Levels

This method is used to calculate volume of earthwork for the elevations of basements, large tanks and borrow pits. In this method the whole area is divided into a number of rectangles or triangles (Fig. 6.13). The levels are taken at corner points before and also after excavation. The depth of excavation at each corner point is measured. Then for each simple figure (rectangle or triangle).

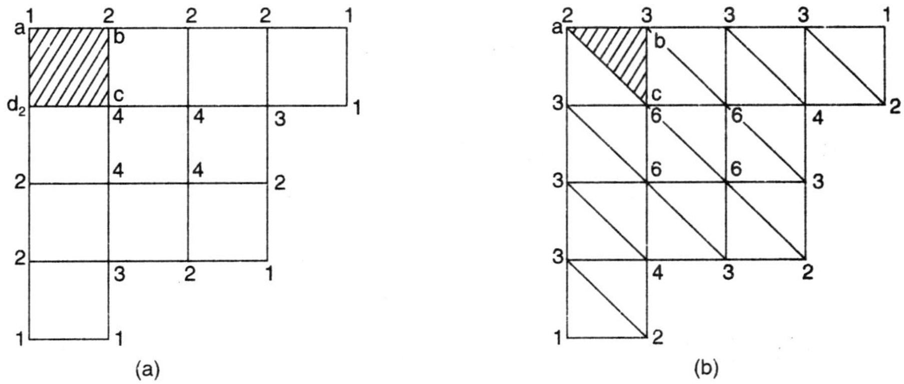

Fig. 6.13

$$V = \text{Area of the figure} \times \text{average depth.}$$

Thus for a rectangle with corner depth h_a, h_b, h_c and h_d,

$$V = \text{Area of rectangle} \times \frac{h_a + h_b + h_c + h_d}{4}$$

For a triangle,

$$V = \text{Area of triangle } \frac{h_a + h_b + h_c}{3}$$

All such volumes, when added give total volume of work.

It may be noted that in Fig. 6.13 (*a*), in total volume calculations depth of some corners appear once, some twice, some of them 3 times and some 4 times. If

Σh_1 = some of depths used once

Σh_2 = sum of depths used twice

Σh_3 = sum of depths used thrice

Σh_4 = sum of depths used four times.

Then

$$V = \frac{A}{4}(\Sigma h_1 + 2\Sigma h_2 + 3\Sigma h_3 + 4\Sigma h_4) \quad \text{... (6.12)}$$

Similarly in Fig. 6.13 (*b*), sum depths are used once, some 2 times, some 3 times and some others 6 times. Defining h_i same as above.

$$V = \frac{A}{3}(\Sigma h_1 + 2\Sigma h_2 + 3\Sigma h_3 + 6\Sigma h_6) \quad \text{... (6.13)}$$

■ **Example 6.8:** *A 60 m × 60 m plot is to be excavated to a formation level of 80.0 m. The present levels at 20 m × 20 m grid are as shown in Fig. 6.14. Calculate the volume of earth work.*

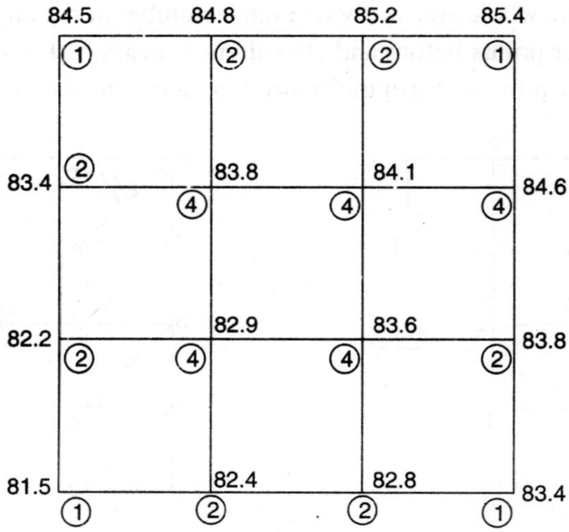

Fig. 6.14

Solution: The number of times a particular corner depth is used in volume calculation is marked in circles. Formation level is 80.0 m.

$\therefore \quad \Sigma h_1 = 4.5 + 5.4 + 3.4 + 1.5 = 14.8$ m

$\Sigma h_2 = 4.8 + 5.2 + 3.4 + 4.6 + 2.2 + 3.8 + 2.4 + 2.8$

$= 29.2$ m

$\Sigma h_3 = 0$

$\Sigma h_4 = 3.8 + 4.1 + 2.9 + 3.6 = 14.4$ m

Area of each grid, A = 20 × 20 = 400 m².

$$\therefore \quad V = \frac{400}{4}\,(14.8 \times 1 + 29.2 \times 2 + 0 \times 3 + 14.4 \times 4)$$

$$= 13080 \text{ m}^3.$$

6.3.3 Computation of Volume from Contours

Figure 6.15 shows a dam with full water level of 100 m and contours on upstream side. Capacity of reservoir to be found is nothing but volume of fill with water level at 100 m. The whole area lying within a contour line is found by planimeter. It may be noted that area to be measured is *not* between two consecutive contour lines.

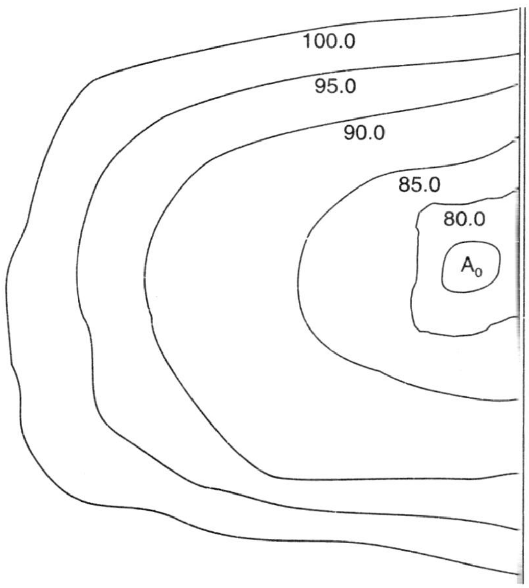

Fig. 6.15

Let $A_0, A_1, A_2, ..., A_n$ be area of contours and h be contour interval. Then from trapezoidal rule:

$$V = h\left[\frac{A_0 + A_n}{2} + A_1 + A_2 + \cdots + A_{n-1}\right]$$

and by prismoidal rule:

$$V = \frac{h}{3}\,[(A_0 + A_n) + 4\,(A_1 + A_3 + \cdots + A_{n-1}) + 2\,(A_2 + A_4 + \cdots + A_{n-2})$$

where there are n segments and n is even number.

■ **Example 6.9:** *The area within the contour lines at the site of reservoir and along the face of the proposed dam are as follows:*

Contour	*Area in m²*
100 m	*800*
104 m	*9600*
108 m	*11800*
112 m	*12400*

116 m	*14300*
120 m	*18400*
124 m	*20360*

Assuming 100 m as the bottom level of the reservoir, and 124 m as full level, calculate the capacity of the reservoir using trapezoidal and prismoidal formula.

Solution: (*a*) **Prismoidal Formula:**

$$V = h\left[\frac{A_0 + A_n}{2} + A_1 + A_2 + \cdots + A_{n-1}\right]$$

$$= 4\left[\frac{800 + 20360}{2} + 9600 + 11800 + 12400 + 14300 + 18400\right]$$

$$= \mathbf{308320\ m^3}$$ **Ans.**

(*b*) **Prismoidal Formula:**

$$V = \frac{h}{3}\,[A_0 + A_n + 4\,(A_1 + A_3 + \cdots + A_{n-1}) + 2\,(A_2 + A_4 + \cdots + A_{n-2})]$$

$$= \frac{4}{3}\,[800 + 20360 + 4\,(9600 + 12400 + 18400) + 2\,(11800 + 14300)]$$

$$= \mathbf{313280\ m^3}$$ **Ans.**

QUESTIONS

1. Find the area of the field from the following notes of cross staff survey, all measurements being in metres.

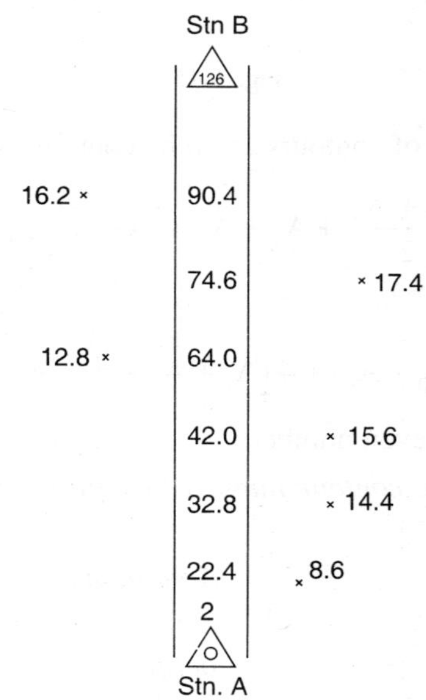

[Ans: 2419.76 m²]

2. The following observations were taken to a boundary from a chain line.

Distance in m	0	10	20	30	40	50	60	70
Offset in m	2.4	3.6	4.2	4.8	4.4	3.8	2.8	1.2

Calculate the area enclosed between the chains line, the boundary line and the end offsets by

(*i*) Trapezoidal rule (*ii*) Simpson's rule **Ans: (*i*) 254.0 m^2 (*ii*) 257.33 m^2**

[Hint: Consider first six segment by Simpson's rule and to this add area of last segment found by Trapezoidal rule].

3. A plot of ground is in the form of a quadrilateral. To find the area the following measurements were taken:

AB = 118.6 m BC = 220.4 m CD = 158.4 m

DA = 340.0 m and AC = 322.0 m

Find the area of the plot. **[Ans: 33384.5 m^2]**

[Hint: Divide quadrilateral into two triangles)

4. Describe the planimeter and explain how it is used for finding the area of a given map.

5. A planimeter was used to measure the area of a map once keeping anchor point outside the figure and second time keeping it inside the figure. The observations are as follows:

(*i*) When anchor point was outside the figure:

Initial reading = 7.364

Final reading = 3.234

The zero of dial passed the index mark once in the clockwise direction.

(*ii*) When anchor point was inside the figure:

Initial reading = 2.384

Final reading = 4.443

The zero of the dial passed the index mark twice in the anticlockwise direction. Take M = 100 cm^2.

[Ans: Area of zero circle = 2381.1 cm^2]

6. How do you determine the multiplying constant and area of zero circle of a planimeter experimentally ?

7. The following are the cross-sectional areas of an embankment at 30 m interval

Distance in m	0	30	60	90	120	150	180
Area in m^2	122	432	612	720	718	1020	1040

Determine the volume of earth work by (*i*) trapezoidal formula (*ii*) prismoidal formula

[Ans: Trapezoidal formula, V = 122490 m^3, Prismoidal Formula, V = 125100 m^3]

8. The areas within the contour lines at the site of a reservoir and along the proposed dam site are as follows:

Contour in m	100	105	110	115	120	125	130
Area in m^2	1250	1420	17300	20200	25200	32400	36780

Assuming 100 m as bottom level of the reservoir and 130 m as full level, determine the capacity of the reservoir using trapezoidal and prismoidal rules.

[Ans: Trapezoidal rule, V = 578675 m^3, Prismoidal rule, V_2 = 565850 m^3]

7

Modern Tools of Surveying and Mapping

Theodolite is an instrument which replaced compass and level. It can measure both horizontal and vertical angles. If telescope is kept at zero reading of vertical angle it serves as an ordinary level. In this modern era of electronics equipments have come up to measure the distances to relieve surveyor from chaining long lines. Total station is another modern survey equipment which combines the features of theodolite and electromagnetic distance measurement (EDM) instruments. Global positioning system is an instrument, which establishes global position of the station making use at least 4 satellite stations. Remote sensing is another revolutionary change in surveying in which objects on the earth are sensed from remote places like aircrafts or satellites and are used for surveying. Geographical Information System (GIS) is a software tool used for the analysis of data of any area using computers. In this spatial data and object data (attribute data) are clubbed and hence information of any place can be easily accessed. In this chapter all these modern tools of surveying and mapping are briefly explained.

7.1 THEODOLITE

It is a commonly used instrument for measuring horizontal and vertical angles. It is used for prolonging a line, levelling and even for measuring the distances indirectly (techeometry). Using verniers angles can be read accurately up to 20″ accuracy. Precise theodolites are available which can read angles up to even 1″ accuracy. They use optical principle for more accurate instruments. Now a days electronic theodolites are also available which display the angles.

In this article construction and use of vernier theodolite is explained.

7.1.1 Parts of a Vernier Theodolite

Figure 7.1 shows a sectional view of a typical vernier theodolite and plate 7.1 shows photograph of such theodolite. Main parts of such a theodolite are:

1. **Telescope:** A telescope is mounted on a horizontal axis (trunnian axis) hence it can rotate in vertical plane. Its length varies from 100 mm 175 mm and its diameter is 38 mm at objective end. Its functions is to provide a line of sight.

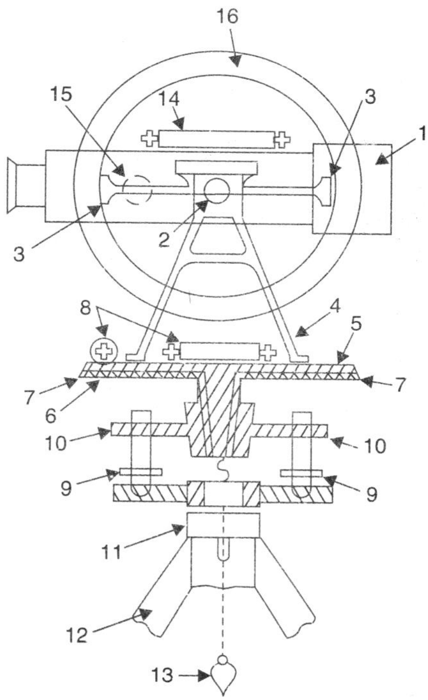

1. Telescope
2. Trunnian axis
3. Vernier frame
4. Standards or A-Frame
5. Upper plate
6. Lower plate
7. Vernier
8. Plate level
9. Foot crews
10. Levelling head
11. Tripod head
12. Tripod
13. Plumb bob
14. Altitude level
15. Focussing screw
16. Vertical circle

Fig. 7.1 Sectional view of a transit theodolite

Plate 7.1 Theodolite

2. **Vertical Circle:** A vertical circle graduated up to an accuracy of 20′ is rigidly connected to the telescope and hence moves with it when the telescope is rotated in vertical plane.

   The graduations are in quadrantal system, 0-0 line being horizontal (Ref. Fig. 7.2).

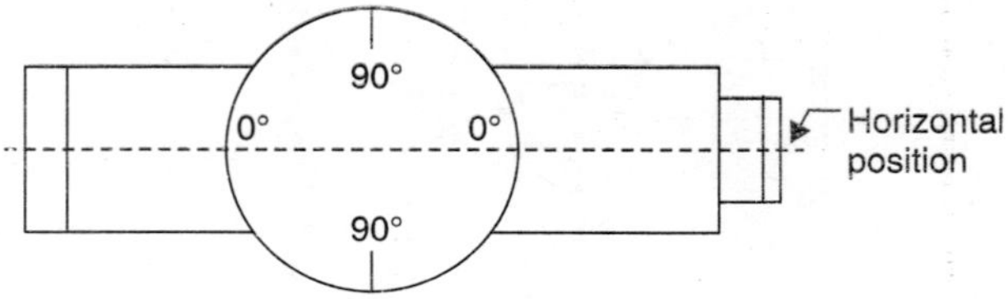

Fig. 7.2

3. **Vernier Frame:** It is a T-shaped frame (Fig. 7.3) consisting of a vertical arm and a horizontal arm. With the help of the climping screws the vertical frame and hence the telescope can be clamped at desired angle. Vertical frame is also known as **T-frame** or **index frame.**

   The vernier arm is known as index arm. At the ends it carries verniers C and D so as to read graduations on vertical circle. They are provided with glass magnifiers. Altitude bubble tube is fitted over the horizontal arm.

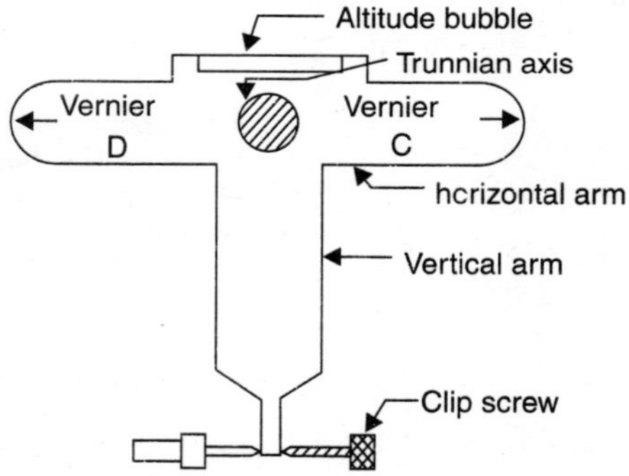

Fig. 7.3

4. **Standards or A-Frame:** The frames supporting telescope are in the form of English letter 'A'. This frame allows telescope to rotate on its trunnian axis in vertical frame. The T-frame and the clamps are also fixed to this frame.
5. **Upper Plate [Fig. 7.4]:** Upper plate supports standards on its top surface. On lower side it is attached to a inner spindle which rotates in the outer spindle of lower plate. Using upper clamp, upper plate can be clamped to lower plate. Using tangent screws, it is possible to give slight relative motion between the two plates, even after clamping. Two diametrically opposite verniers A and B fixed to upper plate help in reading horizontal circle graduations. They are provided with magnifying glasses.

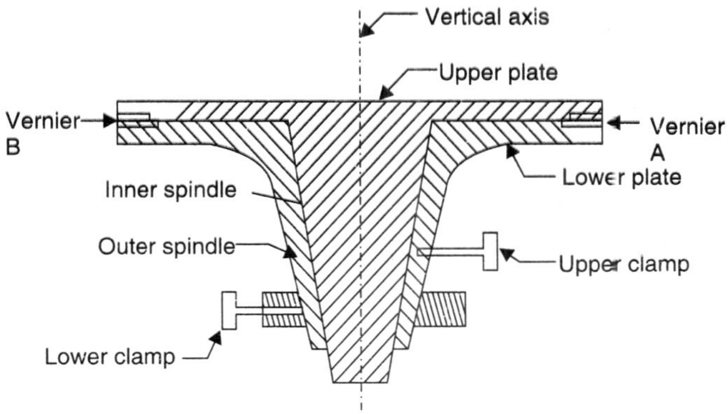

Fig. 7.4

6. **Lower Plate:** The lower plate, attached to the outer spindle carries a graduated circle at its bevelled edge. Graduations are up to an accuracy of 20′. It can be clamped at any desired position using lower clamps. If upper clamp is locked and the lower one is loosened the two plates rotate together. If the upper clamp is loosened and lower clamp locked, upper plate alone rotates. This mechanism is utilised in measuring horizontal angle.
7. **Plate Level:** One or two plate level tubes are mounted on the upper plate. If the two level tubes are provided they will be at right angles to each other one of them being parallel to trunnion axis. These levels help in making the vertical axis of the instrument truely vertical.
8. **Levelling Head:** It consists of two parallel triangular plates known as ***tribratch plates***. The upper tribratch plate is provided with three levelling screws—each one carried by a arm of tribratch plate. By operating screws the levelling of upper plate and hence telescope can be ensured. The lower tribratch can be fitted into a tripod head.
9. **Tripod:** Theodolite is always used by mounting it on a tripod. The legs of tripod may be solid or framed. At the lower end the legs are provided with steel shoes to get good grip with the ground. The top of tripod is provided with external screw to which the lower tribratch plate can be screwed. When not in use tripod head may be protected with a steel cap, provided for this purpose.
10. **Plumb Bob:** A hook is provided at the middle of lower tribratch plate from which a plumb bob can be suspended. It facilitates exact centering of the theodolite on a station.
11. **Shifting Head:** It is provided below the lower plate. In this, one plate slides over another over a small area of about 10 mm radius. The two plates can be tightened in the desired position. It facilitates exact centering of the instruments.
12. **Magnetic Compass:** In some theodolites a magnetic compass is fixed on one of the strands. It is useful if readings are to be recorded with magnetic north as meridian.

7.1.2 Use of Theodolite

Theodolite is used for measuring horizontal and vertical angles. For this the theodolite should be centered on the desired station point, levelled and telescope is focussed. This process of centering, levelling and focussing is called temporary adjustment of the instrument.

Measurement of Horizontal Angle

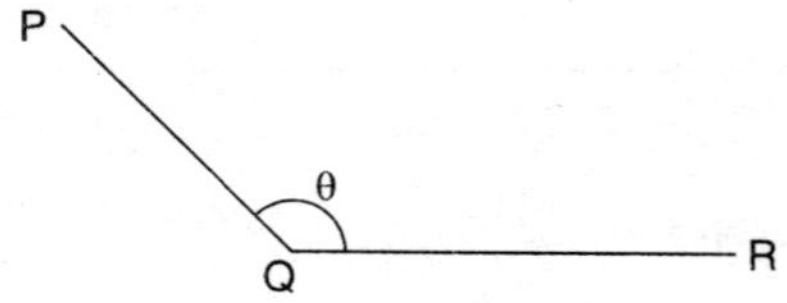

Fig. 7.5

The procedure is explained for measuring horizontal angle θ = PQR at station Q (Ref. Fig. 7.5)

1. Set the theodolite at Q with vertical circle to the left of the line of sight and complete all temporary adjustments.
2. Release both upper and lower clamps and turn upper plate to get 0° on the main scale. Then clamp main screw and using tangent screw get exactly zero reading. At this stage vernier A reads 0° and vernier B reads 180°.
3. Through telescope take line of sight to signal at P and lock the lower clamp. Use tangent screw for exact bisection.
4. Release the upper clamp and swing telescope to bisect signal at R. Lock upper clamp and use tangent screen to get exact bisection of R.
5. Read verniers A and B. The reading of vernier A gives desired angle PQR directly, while 180° is to be subtracted from the reading of vernier B to get the angle PQR.
6. Transit (move by 180° in vertical plane) the telescope to make vertical circle to the right of telescope. Repeat steps 2 to 5 to get two more values for the angle.
7. The average of 4 values found for θ, give the horizontal angle. Two values obtained with face left and two obtained with face right position of vectical circle are called one set of readings.
8. If more precision is required the angle may be measured **repeatedly.** *i.e.,* after step 5, release lower clamp, sight signal at P, then lock lower clamp, release upper clamp and swing the telescope to signal at Q. The reading of vernier A doubles. The angle measured by vernier B is also doubled. Any number of repetitions may be made and average taken. Similar readings are then taken with face right also. Finally average angle is found and is taken as desired angle 'Q'. This is called **method of repetition.**
9. There is another method of getting precise horizontal angles. It is called **method of reiteration.** If a number of angles are to be measured from a station this technique is used (Fig. 7.6). With zero reading of vernier A signal at P is sighted exactly and lower clamp and its tangent screw are locked. Then θ_1 is measured by sighting Q and noted. Then θ_2, θ_3 and θ_4 are measured by unlocking upper clamp and bisecting signals at R, S and P. The angles are calculated and checked to see that sum is 360°. In each case both verniers are read and similar process is carried out by changing the face (face left and face right).

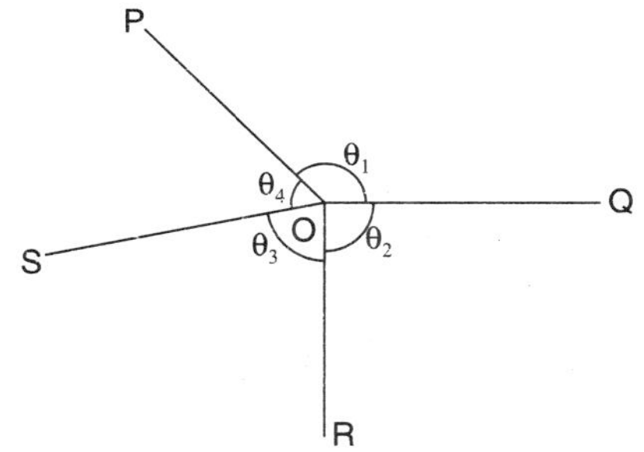

Fig. 7.6

Measurement of Vertical Angle

Horizontal sight is taken as zero vertical angle. Angle of elevations are noted as +ve angles and angle of depression as –ve angles.

To measure vertical angle the following procedure may be followed.

1. Complete all temporary adjustment at the required station.
2. Take up levelling of the instrument with respect to altitude level provided on the A – frame. This levelling process is similar to that used for levelling dumpy level *i.e.,* first altitude level is kept parallel to any two levelling screws and operating those two screws bubble is brought to centre. Then by rotating telescope, level tube is brought at right angles to the original position and is levelled with the third screw. The procedure is repeated till bubble is centred in both positions.
3. Then loosen the vertical circle clamp, bisect P and lock the clamp. Read verniers C and D to get vertical angle. Take the average as the actual vertical angle.

7.2 ELECTROMAGNETIC DISTANCE MEASURING INSTRUMENTS

Sun light or artificially generated electromagnetic waves consists of waves of different lengths. The spectrum of an electromagnetic wave is as shown below:

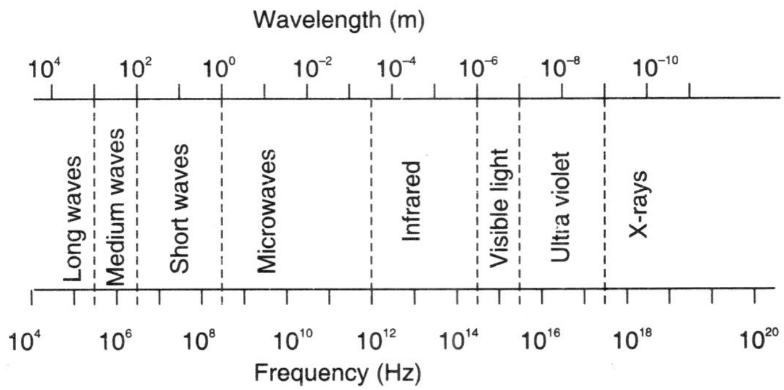

Among these waves microwaves, infrared waves and visible light waves are useful for the distance measurement. In EDM instruments these waves are generated, modulated and then propagated. They are reflected at the point up to which distance is to be measured from the instrument station and again received by the instrument. The time taken by the wave to travel this $2x$ distance may be measured and knowing the velocity of wave, the distance may be calculated. However time is too short, measuring the time taken is difficult. The improved techniques use phase difference method in which the number of completed wave and incomplete wave is measured. Knowing the length of wave, distances are calculated. Built up microprocessors provided in the instrument calculate the distances and display it by liquid crystal display (LCD).

EDM instruments may be classified into the following three types:

1. Micro wave instruments
2. Infra red wave instruments
3. Light wave instruments.

7.2.1 Micro Wave Instruments

These instruments make use of micro waves. Such instruments were invented as early as 1950 in South Africa by Dr. T.L. Wadley and named them as Tellurometers. The instrument needs only 12 to 24 V batteries. Hence they are light and highly portable. Tellurometers can be used in day as well as in night. The range of these instrument is up to 100 km. It consists of two identical units. One unit is used as master unit and the other as remote unit. Just by pressing a button, a master unit can be converted into a remote unit and a remote unit into a master unit. It needs two skilled persons to operate. A speech facility is provided to each operator to interact during measurements.

7.2.2 Infrared Wave Instruments

In these instrument amplitude modulated infrared waves are used. Prism reflectors are used at the end of line to be measured. These instruments are light and economical and can be mounted on theodolite. With these instruments accuracy achieved is ± 10 mm. The range of these instruments is up to 3 km. These instruments are useful for most of the civil engineering works. These instruments are available in the trade names DISTOMAT DI 1000 and DISTOMAT DI 55.

7.2.3 Light Wave Instruments

These instruments rely on propagation of modulated light waves. This type of instrument was first developed in Sweden and was namad as Geodimeter. During night its range is up to 2.5 km while in day its range is up to 3 km. Accuracy of these instruments varies from 0.5 mm to 5 mm/km distance. These instruments are also very useful for civil engineering projects.

The advantage of using EDM instruments is the speed and accuracy in measurement. Several obstacles to chaining are automatically over come when these instruments are used.

7.3 TOTAL STATION

It is combination of EDM instrument and electronic theodolite. It is also integrated with microprocessor, electronic data collector and storage system. The instrument can be used to measure horizontal and vertical angles as well as sloping distance of object to the instrument. Microprocessor unit processes the data collected to compute:

1. average of multiple angles measured
2. average of multiple distance measured.
3. horizontal distance
4. distance between any two points
5. elevation of objects and
6. all the three coordinates of the observed point.

Data collected and processed may be down-loaded to computers for further processing. Total station is a compact instrument and weighs 50 to 55 N. A person can easily carry it to the field. Total stations with different accuracies, in angle measurement and different range of measurements are available in the market. [See plate 7.2 also]. Figure 7.7 shows one such instrument manufactured by SOKKIA Co. Ltd. Tokyo, Japan.

Plate 7.2 Total station

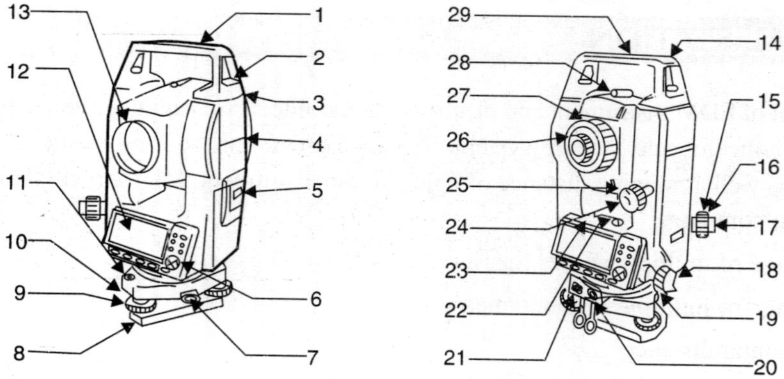

1. Handle
2. Handle securing screw
3. Data input/output terminal (Remove handle to view)
4. Instrument height mark
5. Battery cover
6. Operation panel
7. Tribrach clamp (SET300*S*/500*S*/600S: Shifting clamp)
8. Base plate
9. Levelling foot screw
10. Circular level adjusting screws
11. Circular level
12. Display
13. Objective lens
14. Tubular compass slot
15. Optical plummet focussing ring
16. Optical plummet reticle cover
17. Optical plummet eyepiece
18. Horizontal clamp
19. Horizontal fine motion screw
20. Data input/output connector (Besides the operation panel on SET600/600*S*)
21. External power source connector (Not included on SET600/600*S*)
22. Plate level
23. Plate level adjusting screw
24. Vertical clamp
25. Vertical fine motion screw
26. Telescope eyepiece
27. Telescope focussing ring
28. Peep sight
29. Instrument center mark

Fig. 7.7 Parts of total station

Brief Description of Important Operations

Distance Measurement: Electronic distance measuring (EDM) instrument is a major part of total station. Its range varies from 2.8 km to 4.2 km. The accuracy of measurement varies from 5 mm to 10 mm per km measurement. They are used with automatic target recognisers. The distance measured is always sloping distance from instrument to the object.

Angle Measurements: The electronic theodolite part of total station is used for measuring vertical and horizontal angle. For measurement of horizontal angles any convenient direction may be taken as reference direction. For vertical angle measurement vertical upward (zenith) direction is taken as reference direction. The accuracy of angle measurement varies from 2 to 6 seconds.

Data Processing : This instrument is provided with an inbuilt microprocessor. The microprocessor averages multiple observations. With the help of slope distance and vertical and horizontal angles measured, when height of axis of instrument and targets are supplied, the microprocessor computes the horizontal distance and X, Y, Z coordinates. The processor is capable of applying temperature and pressure corrections to the measurements, if atmospheric temperature and pressures are supplied.

Display: Electronic display unit is capable of displaying various values when respective keys are pressed. The system is capable of displaying horizontal distance, vertical distance, horizontal and vertical angles, difference in elevations of two observed points and all the three coordinates of the observed points.

Electronic Book: Each point data can be stored in an electronic note book (like compact disc). The capacity of electronic note book varies from 2000 points to 4000 points data. Surveyor can unload the data stored in note book to computer and reuse the note book.

Use of Total Station

The instrument is mounted on a tripod and is levelled by operating levelling screws. Within a small range instrument is capable of adjusting itself to the level position. Then vertical and horizontal reference directions are indexed using onboard keys. It is possible to set required units for distance, temperature and pressure (FPS or SI). Surveyor can select measurement mode like fine, coarse, single or repeated. When target is sighted, horizontal and vertical angles as well as sloping distances are measured and by pressing appropriate keys they are recorded along with point number. Heights of instrument and targets can be keyed in after measuring with them tapes. Then processor computes various information about the point and displays on screen. This information is also stored in the electronic note book. At the end of the day or whenever electronic note book is full, the information stored is downloaded to computers.

The point data downloaded to the computer can be used for further processing. There are software like auto civil and auto plotter clubbed with autocad which can be used for plotting contours at any specified interval and for plotting cross-section along any specified line.

Advantages of Using Total Stations

The following are some of the major advantages of using total station over the conventional surveying instruments:

1. Field work is carried out very fast.
2. Accuracy of measurement is high.
3. Manual errors involved in reading and recording are eliminated.
4. Calculation of coordinates is very fast and accurate. Even corrections for temperature and pressure are automatically made.
5. Computers can be employed for map making and plotting contour and cross-sections. Contour intervals and scales can be changed in no time.

However, surveyor should check the working condition of the instruments before using. For this standard points may be located near survey office and before taking out instrument for field work, its working is checked by observing those standard points from the specified instrument station.

7.4 GLOBAL POSITIONING SYSTEM

The station points used in surveying are to be identified before executing any project: For this purpose, surveyors used permanent objects as reference points and made reference sketches of station points.

Navigators used sun and stars as references. Sometimes when the project is taken up the so called parmanent object (like building corner) may not exist when the execution of project work is taken up. For navigations weather conditions may obstruct the observations. Now a days this problem is over come by using an instrument called Global Positioning System (GPS). This was developed by United States defence department and was called as Navigational System with Time and Ranging Global Positioning System (NAVSTAR) or which is now simply known as GPS.

There are 24 geostationary satellites positioned around the earth by US air force. These satellites are used as reference points to locate any point on the earth. These satellites are at an altitude of 20200 km above the earth. The 24 satellites are positioned such that from any point on the earth a minimum of 4 satellites are visible.

A user needs only GPS receiver. The receiver measures the travel time of the signals from satellites and calculate position (latitude and longitude) and the elevation altitude of the station with reference to a selected datum. The advantages of using GPS are:

1. Can be used in day as well as in night.
2. Intervisibility of the two stations on the earth is not a requirement.
3. Time required to establish the position of a point is much less.
4. Man power required is less.
5. Accuracy is high. Most expensive GPS provide accuracies within 10 mm.

Uses of GPS

GPS is very useful in

1. Marine navigation
2. Airborne aviation
3. Surveying of land.
4. Sports such as yatching, hiking.
5. The sophistication of GPS has improved so much that drivers of automobiles can get directions to their destinations easily on the screen.

7.5 REMOTE SENSING

Remote sensing may be defined as art and science of collecting informations about objects, area or phenomenon without having physical contact with it. Eye sight and photographs are common examples of remote sensing in which sunlight or artificial light energy from electricity is made to strike the object. Light energy consists of electromagnetic waves of all lengths and intensity. When electromagnetic wave falls on the object, it is partly

1. absorbed
2. scattered
3. transmitted
4. reflected.

Different objects have different properties of absorbing, scattering, transmitting and reflecting the energy. By capturing reflected waves with sensors, it is possible to identify the objects. However this remote sensing has its own limitations in terms of distance and coverage of area at a time.

Photographic survey, in which photographs taken from aircrafts are used for map making, fall under this category of remote sensing. Using electronic equipments, this basic remote sensing technique is extended to identifying and quantifying various objects on the earth by observing them from longer distances from the space. For this purpose, geostationary satellites are launched in the space, which rotate around the earth at the same speed as earth. Hence the relative velocity is zero and they appear stationary when observed from any point on the earth. Depending upon the property of the object, the electromagnetic waves sent from the satellite reflected energy is different. The reflected waves in the band width of infrared, thermal infrared and micro waves are picked up by sensors mounted on satellite. Since each feature on the earth has different reflection property, it is possible to identify the features on the earth with satellite pictures. Data obtained from satellites are transferred to ground stations through RADARS where user analyses to find out the type of object and the extent of it. This is called image processing. For quantifying the objects computers are used. India is having its own remote sensing satellites like IRS-series, INSAT series and PSLV series.

Application of Remote Sensing

Various applications of remote sensing may be grouped into the following:

1. Resource exploration
2. Environmental study
3. Land use
4. Site investigation
5. Archaeological investigation and
6. Natural hazards study.

1. *Resource Exploration*: Geologists use remote sensing to study the formation of sedimentary rocks and identify deposits of various minerals, detect oil fields and identify underground storage of water. Remote sensing is used for identifying potential fishing zone, coral reef mapping and to find other wealth from ocean.
2. *Environmental Study*: Remote sensing is used to study cloud motion and predict rains. With satellite data it is possible to study water discharge from various industries to find out dispersion and harmful effects, if any, on living animals. Oil spillage and oil slicks can be studied using remote sensing.
3. *Land Use:* By remote sensing, mapping of larger areas is possible in short time. Forest area, agricultural area, residential and industrial area can be measured regularly and monitored. It is possible to find out areas of different crops.
4. *Site Investigation*: Remote sensing is used extensively in site investigations for dams, bridges, pipe lines. It can be used to locate construction materials like sand and gravel for the new projects.
5. *Archaeological Investigation*: Many structures of old era are now buried under the ground and are not known. But by studying changes in moisture content and other characteristics of the buried objects and upper new layer, remote sensors are able to recognise the buried structutures of archaeological importance.
6. *Natural Hazard Study*: Using remote sensing the following natural hazards can be predicted to some extent and hazards minimised:

1. Earthquake
2. Volcanoes
3. Landslides
4. Floods and
5. Hurricane and cyclones.

7.6 GEOGRAPHICAL INFORMATION SYSTEM (GIS)

Maps are used as the languages of simple geography. Importance of map making is recognised long ago. Surveyors went round the land and prepared maps. Data required for locating and calculating extent of a place/region is called spatial data.

Physical properties and human activities related to a place/region are stored in the form of tables, charts and texts. This information is called attribute data.

Referring to maps/plans and then to attribute data stored in hard copies like books is time consuming updating and managing the data is difficult.

This problem is overcome by combining spatial data and attribute data of the location by appropriate data base management in computers. The location information (spatial data) is digitised from available maps and stored in computers. For this data structure used is either raster data or vector data format. In raster data structures pickcells are associated with the spatial information, while in vector data structure coordinates are associated with each region and sub-regions. Over the spatial data attribute data is overlayed and stored. Once this geographical information system is developed, the user can access the attribute data of any place by clicking over the spatial data of that place. The user can utilised the information for further analysis, planning or for the management. For example, if land records of a village is developed as GIS data, the user can click the state map to pick up the district map and then access taluka map. Then he will access it to pick up the village map. Then land record of that village can be obtained and property map of any owner can be checked and printed. All this can be achieved in a very short time from any convenient place.

Remote sensing and GIS go hand in hand, since lot of data for GIS is from remote sensing. Remote sensing needs GIS for data analysis. Some of the areas of GIS application are:

1. drainage systems
2. streams and river basins management
3. lakes
4. canals
5. roads
6. railways
7. land records
8. layout of residential areas
9. location of market, industrial, cultural and other utilities
10. land use of different crops etc.

The above information helps in planning infrastructural development activities such as planning roads, rail routes, dams, canals, tunnels, etc. It helps in taking steps to check hazards of soil erosion and environmental pollution. Monitoring of crop pattern and condition helps in taking necessary action to the challenges in future.

QUESTIONS

1. How do you measure a horizontal angle with a theodolite?
2. What is face right and face left readings in theodolite survey?
3. What is meant by method of repetation and method of reiteration in theodolite survey?
4. Write short notes on the following:
   (*a*) EDM instruments
   (*b*) Total station
   (*c*) Global positioning system
   (*d*) Remote sensing
   (*e*) Geographical Information System.
5. What are the advantages of using total station over using conventional instruments in surveying?
6. List various areas of application of remote sensing and GIS.

8

Construction Materials

Several materials are required for construction works. Traditional materials like stone, bricks, sand, aggregates, lime and timbers are being used as construction materials for centuries. Cement, mortar, concrete and ceramic materials and steel brought major changes in construction activities. Paints, varnishes, glasses and plastics and many miscellaneous materials have added aesthetic aspect to buildings. In this chapter requirement, types, uses and properties of some of these materials is presented.

8.1 STONE

Stone is a naturally available building material, which has been used from the early age of civilisation. It is available in the form of rocks, which is cut to required size and shape and used as building block. It has been used to build small residential buildings to large palaces and temples all over the world. Red Fort, Taj Mahal and Vidhan Sabha at Bangalore are famous stone buildings.

Requirement of a Good Building Stone

The requirements of good building stones for construction works are given below:

1. **Strength:** A good building stone should be able to resist the load coming on it. Ordinarily this is not of primary importance since all stones are having good strength. However in case of large structures, it may be necessary to check the strength.
2. **Durability:** Stone selected should be capable of resisting adverse effects of natural forces like wind, rain, varying temperature conditions etc.
3. **Hardness:** The stones used in floors, pavements should be able to resist abrasive forces caused by movement of men and materials over them.
4. **Toughness:** Building stones should be tough enough to sustain stresses developed due to vibrations of machinery and moving loads over them. The stone aggregates used for road construction should be hard as well as tough.
5. **Specific Gravity:** Havier variety of stones should be used for the construction of dams, retaining walls, docks and harbours. The specific gravity of good building stone is between 2.4 and 2.8.

6. **Porosity and Absorption:** A building stone should not be porous. If it is porous, rain water enters into the pour and reacts with stone and crumbles it. In higher altitudes, freezing of water in pores of stone disintegrates the stone.
7. **Dressing:** Giving required shape to the stone is called dressing. It should be easy to dress so that cost of dressing is reduced. However care should be taken that, this cannot be at the cost of required strength and durability.
8. **Appearance:** In case of stones to be used for face works, where appearance is a primary requirement, its colour and ability to receive polish is an important factor.
9. **Seasoning:** Good stones should be free from quarry sap. Lateritic stones are not used immediately after quarrying. They are allowed to get rid of quarry sap for 6 to 12 months. This process is known as seasoning.
10. **Cost:** Cost is an important consideration in selecting a building material. Proximity of the quarry to building site brings down the cost of transportation and hence the cost of the stones.

Types of Stones

Based on their origin of formation, building stones are classified into three main groups—Igneous, sedimentary and metamorphic rocks.

Igneous Rocks: These rocks are formed by cooling and solidifying of the rock masses from their molten magmatic condition of the material of the earth. Generally igneous rocks are strong and durable. Granite, trap and basalt are the rocks belonging to this category. Granites are formed by slow cooling of the lava under thick cover on the top. Hence they have crystalline surface. The cooling of lava at the top surface of earth results into non-crystalline and glassy texture. Trap and basalt belong to this category.

Sedimentary Rocks: Sedimentary rocks are formed by flowing water depositing its suspended materials at some points of obstacles to its flow. These rocks are more uniform, fine grained and compact in their texture. They represent a bedded or stratified structure in general. Sand stone, lime stones, mud stone etc. belong to this class of rock.

Metamorphic Rocks: These are formed by the metamorphic action of pressure and internal heat or both on previously formed igneous and sedimentary rocks. As a result of metamorphic action granites become gneisses, trap and basalt change to chists and laterite, limestone changes to marble. Sand stone becomes quartzite and mud stone becomes slate.

Uses of Stones in Construction

1. Stone masonry is used for foundations, walls columns and arches.
2. Stones are used for flooring.
3. Stone slabs are used as damp proof courses, lintels and even as roofing materials.
4. Stones with good appearance are used for the face works of buildings. Polished marbles and granite are commonly used materials.
5. Stones are used for paving of roads, foot path and open spaces around the buildings.
6. Stones are also used in the construction of piers and abutments of bridges, dams and retaining walls.

7. Crushed stones with murum are used to provide base course for roads. When mixed with tar they form finishing coat.
8. Crushed stones are also used for the following works:
   (*a*) As a basic inert material in concrete
   (*b*) For making artificial stones and building blocks
   (*c*) As railway ballast.

Properties and Importance of Stones as Construction Material

There is not a single stone which fulfils all the requirement of a good building stone, since one requirement may contradict another. For example strength and durability contradicts ease of dressing. Hence it is necessary for site engineer to look into the properties required for the intended work and select the stone. The following properties of the stones should be looked into before selecting stones for engineering works:

1. *Structure:* Structure of the stone may be stratified (layered) or unstratified. Structured stones can be easily dressed and hence suitable for super structure. Unstratified stones are hard and difficult to dress. They are preferred for the foundation work.

Texture: Fine grained stones with homogeneous distribution look attractive and hence they are used for carving. Such stones are usually strong and durable.

Density: Light weight stones are weak. Stones with specific gravity less than 2.4 are considered unsuitable for building.

Appearance: A stone with uniform and attractive colour is durable if grains are compact. Marble and granite get very good appearance, when polished. Hence they are used for face works in buildings.

Strength: Indian standard code recommends a minimum crushing strength of 3.5 N/mm^2 for any building block. Table 8.1 shows the crushing strength of various stones. Due to non-uniformity of the material, usually a factor of safety of 10 is used to find the permissible stress in a stone. Hence even laterite can be used safely for a single storey building, since in such structure expected load can hardly give a stress of 0.15 N/mm^2. However care should be taken to check the stresses, if beam is placed over laterite wall.

Table 8.1

Name of stone	*Crushing strength in N/mm^2*
Trap	300 to 350
Basalt	153 to 189
Granite	104 to 140
Slate	70 to 210
Marble	72
Sand stone	65
Lime stone	55
Laterite	1.8 to 3.2

Hardness: It is an important property to be considered when stone is used for flooring and pavement. Coefficient of hardness is to be found by conducting test on standard specimen in Dory's testing machine as:

$$= 20 - \frac{1}{3} \times \text{loss in weight in grams}$$

For road works coefficient of hardness should be at least 17. Stone with coefficient of hardness less than 14 should not be used for any building work.

Percentage Wear: It is measured by attrition test. It is an important property to be considered in selecting stone aggregates for road work and railway ballast. A good stone should not show wear of more than 2%.

Porosity and Absorption: All stones have pores and hence absorb water. The reaction of water with the material of stone cause disintegration. Absorption test is specified as percentage of water absorbed by the stone when it is immersed under water for 24 hours. For a good stone should be as small as possible and in no case more than 5.

Weathering: Rain and wind cause loss of good appearance of stones. Hence stones with good resistance to weathering should be used for face works.

Toughness: The resistance to impact is called taughness. It is determined by impact test stones with toughness index more than 19 are preferred for road work. Toughness index 13 to 19 are considered as medium tough and those with toughness index less than 13 are poor stones.

Resistance to Fire: Sand stone resists fire better. Argillaceous (clay) materials, though poor in strength, are good in resisting fire.

Ease in Dressing: Cost of dressing contributes to cost of stone masonry to a great extent. Dressing is easy in stones of lesser strength. Hence an engineer should look into sufficient strength rather than high strength while selecting stones for building works.

Seasoning: The stones attained from quarry contain moisture in the pores. The strength of stone improves if this moisture is removed before using the stone. The process of removing moisture from pores is called seasoning. The best way of seasoning is to allow it to the action of nature for 6 to 12 months. This is very much required in case of laterite stones.

Thus stone is an important material in a number of civil engineering works.

8.2 BRICK

Brick is obtained by moulding good clay into a block, which is dried and then burnt. This is the oldest building block to replace stone. Manufacture of brick started with hand moulding, sun drying and burning in clamps. A considerable amount of technological development has taken place with better knowledge of properties of raw materials, better machineries and techniques of handling, drying and burning.

The size of the brick is maintained such that a mason can handle them with one hand. Indian standard bricks are of size 190 mm × 90 mm × 90 mm and 190 mm × 90 mm × 40 mm. With mortar joints, the size of these bricks are taken as 200 mm × 100 mm × 100 mm and 200 mm × 100 mm × 50 mm. However the old size of $8\frac{3}{4}'' \times 4\frac{1}{4}'' \times 2\frac{5}{8}''$ giving a final size of $9'' \times 4\frac{1}{2}'' \times 3''$ is still commonly used in India.

Requirement and Properties of Good Building Bricks

Colour: Colour should be uniform and bright.

Shape: Bricks should have plane faces. They should have sharp and true right angled corners.

Size: Bricks should be of standard size as prescribed by code.

Texture: Brick should possess fine, dense and uniform texture. It should not possess fissures. cavities, loose grit and unburnt time.

Soundness: When struck with hammer or with another brick, it should give metallic sound.

Hardness: Finger scratching should not leave any impression on the brick.

Strength: Crushing strength of brick should not be less than 3.5 N/mm^2. A field test for strength is that when dropped from a height of 0.9 m to 1.0 m on a hard ground, the brick should not break into pieces.

Water Absorption: After immercing brick in water for 24 hours, water absorption should not be more than 20 per cent by weight. For class-I works, this limit is 15 per cent.

Efflorescence: Bricks should not show white patches when soaked in water for 24 hours and then allowed to dry in shade. White patches are due to the presence of sulphates of calcium, magnesium and potassium. They keep the masonry permanently in damp and wet conditions.

Thermal Conductivity: Bricks should have low thermal conductivity, so that buildings built with them are cool in summer and warm in winter.

Sound Insulation: Havier bricks are poor insulators of sound while light weight and hollow bricks provide good sound insulation.

Fire Resistance: Fire resistance of bricks is usually good. In fact bricks are used to encase steel columns to protect them from fire.

Types of Bricks

Bricks may be broadly classified as

1. Building bricks
2. Paving bricks
3. Fire bricks and
4. Special bricks.

1. Building Bricks: These include common building bricks, used for the construction of walls.

2. Paving Bricks: These are vitrified bricks used as pavers.

3. Fire Bricks: These include bricks specially made to withstand furnace temperatures, such as silica bricks.

4. Special Bricks: These bricks are different from the commonly used building bricks with respect to their shape and purpose for which they are made: Some of such bricks are listed below:

(*a*) Specially shaped bricks (*b*) Facing bricks

(*c*) Perforated building bricks (*d*) Burnt clay hallow bricks

(*e*) Sewer bricks (*f*) Acid resistance bricks.

(*a*) **Specially Shaped Bricks:** Bricks of special shapes are manufactured to meet the requirements of different situations. Some of them are shown in Fig. 8.1

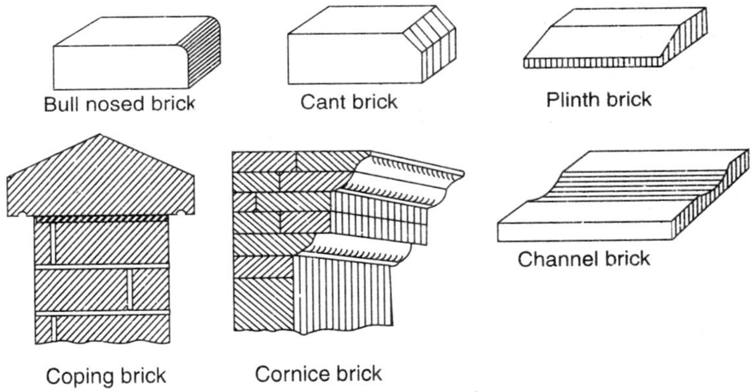

Fig. 8.1 Special shaped bricks

(*b*) **Facing Bricks:** These bricks are used in the outer face of masonry. Once these bricks are provided, plastering is not required. For these bricks efflorescence requirement should be nil. The standard size of these bricks are 190 × 90 × 90 mm or 190 × 90 × 40 mm.

(*c*) **Perforated Building Bricks:** Bricks are manufactured with area of performance of 30 to 45 per cent. The area of each perforation should not exceed 500 mm^2. The perforation should be uniformly distributed over the surface. They are manufactured in the size 190 × 190 × 90 mm and 290 × 90 × 90 mm size.

(*d*) **Burnt Clay Hollow Bricks:** Figure 8.2 shows a burnt clay hollow brick. They are light in weight. They are used as partition walls. They provide good thermal insulation to building. They are manufactured in the sizes 190 × 190 × 90 mm, 290 × 90 × 90 mm and 290 × 140 × 90 mm. The thickness of any shell should not be less than 11 mm and that of any web not less than 8 mm. The crushing strength should not be less than 3.5 N/mm^2 and absorption should not be more than 20 per cent.

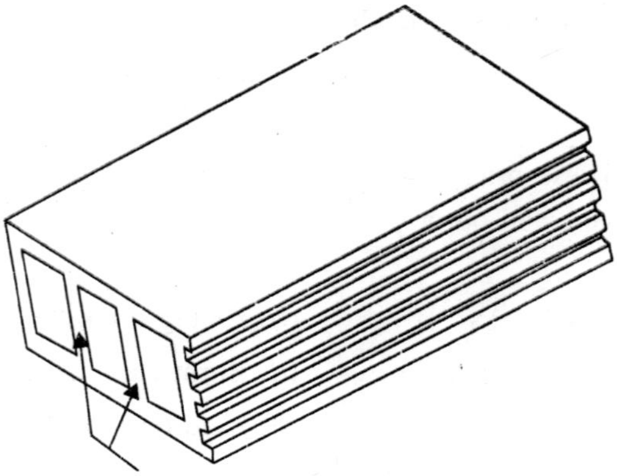

Fig. 8.2 Hollow bricks

(*e*) **Sewer Bricks:** These bricks are meant for construction of sewage lines. These bricks are manufactured from surface clay, fire clay shale or with combination or these. They are available in sizes 190 × 90 × 90 mm and 190 × 90 × 40 mm size. The average strength of these bricks should be a manimum of 17.5 N/mm^2. The water absorption should not be more than 10 per cent.

(*f*) **Acid Resistant Bricks:** These bricks are used for flooring subjected to acid attacks, lining of chambers in chemical plants, lining of sewers carrying industrial wastes etc. These bricks are made of clay or shale of suitable composition with low lime and iron content, flint or sand and vitrified at high temperature in a ceramic kiln.

Uses of Bricks

1. Bricks are extensively used as building blocks.
2. They are used for lining of ovens, furnaces and chimneys.
3. Fire protection to steel columns is provided by encasing them with bricks.
4. Brick bats are used to provided water proofing course to R.C.C. roofs.
5. Foot paths and cycle tracks in many cities are made using bricks.
6. Bricks are used for lining sewer lines.

8.3 LIME

It is an important binding material used in building construction. Lime has been used as material of construction from ancient time. When it is mixed with sand it provides lime mortar and when mixed with sand and coarse aggregates it forms lime concrete.

Classification of Limes, Their Properties

The limes which are obtained by calcination of lime stones are broadly classified as:

1. Fat lime
2. Hydraulic lime and
3. Poor lime.

1. Fat Lime: It is composed of about 95 percentage of calcium oxide. When water is added it slake vigorously and its volume increases to about 2 to $2\frac{1}{2}$ times. Its colour is white.

Its properties are:

(*a*) It hardens slowly
(*b*) It has high degree of plasticity
(*c*) It sets slowly in the presence of air
(*d*) Its colour is white
(*e*) It slakes vigorously.

It is used for

(*a*) White washing and plastering.
(*b*) Its mortar is used for making thin joints of brick work.
(*c*) With surkhi (powder obtained by girding burnt bricks) this mortar is useful for thick masonry walls and foundations.

2. Hydraulic Lime: It contains clay and ferrous oxide. Depending upon the percentage of clay, the hydraulic lime is divided into the following three types:

(*a*) Feebly hydraulic lime (5 to 10% clay content)
(*b*) Moderately hydraulic lime (11 to 20% clay content)
(*c*) Emiently hydraulic lime (21 to 30% clay content).

Hydraulic lime sets under water and hence the name. The colour of hydraulic lime is not perfectly white. It forms a thin paste with water and will not dissolve in water. Hydraulic lime ground in fine powder and mixed with sand is usually kept in the form of heap for a week. Then it is ground again and used for plaster work.

3. Poor Lime: It contains more than 30% clay. Its colour is muddy and it has poor binding property. The mortar made with such lime is used for inferior works.

IS 712 – 1973 classifies lime as class A, B, C, D and E.

Uses of Lime

The following are the uses of lime in construction works:

1. It is used for white washing and to serve as base coat for distemper.
2. It is used for making mortar for masonry work.
3. Lime mortar is used for plastering the walls.
4. It is used for the production of lime-sand bricks.
5. It is used for soil stabilisation.
6. It is used as a refractory material for lining open – hearth furnaces.

8.4 CEMENT

Cement is a commonly used binding material in the construction. The natural cement was obtained by burning and crushing the stones containing clay, carbonate of lime and some amount of carbonate of magnesia. The artificial cement is obtained by burning a mixture of calcarious (calcium) and argillaceous (clay) material at a very high temperature and then grinding the clinker so produced to fine powder. It was first produced in 1824 by a mason Joseph Aspdin in England who patented it as Portland cement.

The requirement of cement are

1. It should act as good binding material
2. It should be easily workable
3. It should offer good resistance to moisture
4. It should possess good plasticity
5. It should harden early and
6. It should impart strength to masonry.

Compounds in Cement

Portland cement consists of the following chemical compounds:

1. Tricalcium silicate	$3CaO.SiO_2$ (C_3S)	40%
2. Dicalcium silicate	$2CaO.SiO_2$ (C_2S)	30%
3. Tricalcium aluminate	$3CaO.Al_2O_3$ (C_3A)	11%
4. Tetracalcium aluminate	$4CaO.Al_2O_3Fe_2O_3$ (C_3AF)	11%

There are also impurities present such as free calcium oxide (CaO) and magnesium oxide (MgO).

When water is added C_3A is the first to react and cause initial set. It generates great amount of heat. C_3S hydrates early and develops strength in first 28 days. It also generates heat. C_2S is the next to hydrate. It hydrates slowly and responsible for ultimate strength. C_4AF is comparatively inactive compound.

Physical Properties of Portland Cement

The following physical properties are to be checked before selecting a cement for civil engineering work. IS 269–1967 specifies the method of testing and prescribes the limits:

1. Fineness
2. Setting time
3. Soundness
4. Crushing strength.

1. Fineness: It is measured in terms of (*i*) percentage of weight retained after sieving cement through 90 micron sieve or (*ii*) surface area of cement in square centimeters per gramme of cement. According to IS code specification the percentage weight retained after sieving through 90 micron sieve should not be more than 10 per cent. In terms of specific surface area of cement per gram as fund by IS code method of test should not be less than 2250 cm^2/gm.

2. Setting Time: A period of 30 minutes as minimum setting time for initial setting and a period of 600 minutes as maximum setting time is specified by IS code, provided the tests are conducted as per IS procedure.

3. Soundness: Once the concrete in structure has hardened, it is necessary to ensure that no volumetric changes takes place. The cement is said to be unsound if it exhibits such volumetric instability. For testing this property IS code recommends test with Le Chatelier mould. When tested as specified in IS 269–1976, the indicator of the Le Chatelier mould should not expand by more than 10 mm.

4. Crushing Strength: For this cement mortar with standard sand are made and tested in compression machines as specified by IS code. For ordinary Portland cement the minimum strength after 3 days should be 16 N/mm^2 and after 7 days it should be 22 N/mm^2.

Types of Cement

In addition to ordinary portland cement there are many varieties of cement. Important varieties are briefly discussed below:

1. **White cement:** The cement when made free from colouring oxides of iron, manganese and chromium results into white cement. For burning of this cement, the oil fuel is used instead of coal. White cement is used for floor finishes, plastering works, ornamental works etc. In swimming pools white cement is used to replace glazed tiles. It is used for fixing marble and glazed tiles.
2. **Coloured cement:** The cements of desired colours are produced by intimately mixing mineral pigments with ordinary cement. The chromium oxide gives green colour. Cobalt produce blue colour. The iron oxide with different proportion produce brown red or yellow colour. Manganese dioxide give black or brown coloured cement. These cements are used for giving finishing of floors, walls, window sills, roofs etc.
3. **Quick setting cement:** By reducing the percentage of gypsum and adding a small percentage of aluminium sulphate and grinding cement finely, quick setting cement is produced. This cement starts setting within 5 minutes after adding of water and becomes hard mass within 30 minutes. This cement is used to lay concrete under static or slowly running water.
4. **Rapid hardening cement:** By burning at high temperatures, increasing lime content and grinding very fine rapid hardening cement can be produced. Though initial and final setting time of this cement are the same as that of portland cement, it gains strength in early days.

   This cement needs shorter curing period and permits earlier removal of forms work. The structure can be put to use early or further construction of the structure can be taken up early.
5. **Low heat cement:** In mass concrete works (like construction of dam), the heat produced due to hydration of cement will not get dispersed easily. It may give rise to cracks. Hence it is preferable to use low heat cement. This cement contains low percentage of C_3A (only 5%) and higher percentage of C_2S (46%).

6. **Pozzuolana cement:** Pozzuolana is a volcanic powder found in Italy. It can be processed from shales and certain types of clays also. In this cement pozzuolana material is 10 to 30 per cent. It can resist action of sulphate and it releases less heat during setting . It imparts higher degree of water tightness. Its tensile strength is high but compressive strength is low. It is used for mass concrete work. It is also used in sewage works.
7. **Expanding cement:** This cement expands as it sets while other cements contract. This is achieved by adding expanding medium like sulpho-aluminate and a stabilising agent to ordinary cement. This is used for filling the cracks in concrete structures.
8. **High alumina cement:** It is manufactured by calcining a mixture of lime and bauxite. It is more resistant to attack by sulphate and acid also to the action of frost. It contains aluminates between 35 to 55 per cent. It develops almost full strength within 24 hours of adding water. It is used for works under water.
9. **Blast furnace cement:** In the manufacture of pig iron, slag comes out as a waste product of blast furnace. By grinding clinkers of cement with about 60 to 65 per cent of slag, this cement is produced. The properties of this cement are more or less same as that of ordinary cement, but it is cheap, since it utilises waste product. This cement is durable but it needs longer period of curing since it gains strength slowly.
10. **Acid resistant cement:** This cement is produced by adding acid resistant aggregates such as quartz, quartzites, sodium fluosilicate, solution of sodium silicate or soluble glass. This cement has good resistance to acid and water. It is commonly used in the construction of chemical factories.
11. **Sulphate-resistant cement:** By keeping the percentage of tricalcium aluminate C_3A below 5 per cent this cement is produced. It is used in the construction of structures which are likely to be damaged by alkaline conditions. Examples of such structures are canals, culverts etc.
12. **Fly ash blended cement:** Fly ash is a by-product in thermal power stations. They contain silica and other chemicals. Particles of fly ash are very small and they fly in air, creating air pollution problem. Thermal power stations spend lot of money to arrest fly ash and dispose. It is found that one of the best way to dispose it is to mix it with cement in controlled condition and derive some of the beneficiary effects on cement. Nowadays cement factories produce the fly ash in their own thermal power stations in controlled conditions or if they borrow it from other thermal stations they further precess the fly ash to make it suitable to blend with cement, 20 to 30 per cent of fly ash is used for blending.

Fly ash blended cements have superior quality of resistance to weather. Hence concrete produced with such cement is more durable. The ultimate strength gained is same as that with ordinary portland cement. However strength gained in the initial stage is slow.

Birla plus, Birla star, A.C.C. Suraksha, silicate cement are some of the brand names of fly ash blended cement in India.

Uses of Cement

Cement is used widely for the construction of various structures. Some of them are listed below:

1. Cement slurry is used for filling cracks in concrete structures.

2. Cement mortar is used for masonry work, plastering, pointing, etc.
3. Cement concrete is used for the construction of various structures like buildings, bridges, water tanks, tunnels, docks, etc.
4. It is useful in the manufacture of lamp posts, telephone posts, railway sleepers piles, etc.
5. It is useful in the manufacture of cement pipes, garden seats, dust bins, flower pots, etc.
6. It is useful for the construction of roads, foot paths, courts for various sports, etc.

8.5 FERROUS AND NON-FERROUS MATERIALS

A ferrous material is one in which iron is a main constituent whereas a non-ferrous material is a metal in which iron is not principal constituent.

8.5.1 Ferrous Materials

Iron are is first converted into pig iron and then pig iron is subjected to various metallurgical processes to mix different percentage of carbon and to get the following three useful ferrous materials:

(*i*) Cast iron —carbon content 1.7–4.5%

(*ii*) Wrought iron—carbon content 0.05–0.15%

(*iii*) Steel—carbon content 0.25–0.25%

All ferrous materials contain about 0.5 to 3% silica, less than 2% manganese, 0.15% sulphur and 0.6% phosphorous.

8.5.2 Cast Iron

Important properties of cast iron are:

1. Compression strength is 700 N/mm^2 and tensile strength is 150 N/mm^2.
2. It is brittle and does not absorb shocks.
3. Its specific gravity is 7.5.
4. Its structure is coarse, crystalline and fibrous.
5. It cannot be magnetised.
6. It does not rust easily.
7. It has low melting point of about 1200°C.

Uses of Cast Iron

1. Used for making rain water and sanitary pipes, sanitary fittings and man hole covers.
2. For railings and spiral stair cases.
3. For fire gratings, brackets for some parts of motors and pumps.

8.5.2 Wrought Iron

It is almost pure iron. Carbon content is less than 0.15% and attempts are made to reduce the other impurities, by oxidisation.

Properties of Wrought Iron

1. Ultimate compressive strength is 200 N/mm^2 and ultimate tensile strength is 375 N/mm^2.
2. It is ductile and malleable.
3. Unit weight is 77 kN/m^3.
4. It melts at about 1500°C. It becomes so soft at 900°C that two pieces can be joined by hammering.
5. It absorbs shocks very well.
6. It forms temporary magnets but it cannot be magnetised permanently.
7. It rusts more easily.

Uses

1. It is used for making nails, nuts and bolts, wires and chains.
2. It is used for making roofing sheets grills, fences, window gaurds, etc.

8.5.3 Steel

It is a extensively used building material. The following three variety of steel are extensively used:

(*a*) Mild steel (*b*) High carbon steel

(*c*) High tensile steel.

(*a*) *Mild steel:* It contains a maximum of 0.25% carbon, 0.055% of sulphur and 0.55% phosphorus.

Properties of steel:

1. It is malleable and ductile.
2. It is more elastic.
3. It can be magnetized permanently.
4. Its specific gravity is 7.8.
5. Its young's modulus is 2.1×10^5 N/mm^2.
6. It can be easily welded.
7. It is equally strong in tension and compression.

Uses

1. Round bars are extensively used as reinforcement in reinforced concrete works.
2. Rolled sections like I, T, L, C, plates etc. are used to build steel columns, beams, trusses, etc.
3. Mild steel tubular sections are also available which are used in structures.

4. Plain and corrugated mild steel are used as roofing materials.
5. Mild steel sections are used in making parts of many machineries.

(*b*) *High carbon steel:* In these carbon percentage is 0.7 to 1.5%.

The properties of high carbon steel are:

1. It is more tough and elastic than mild steel.
2. Welding is more difficult.
3. It can be magnetized permanently.
4. It is stronger in compression than in tension.
5. It withstands shocks and vibrations better.

Uses

1. It is used for making tools such as drills, files, chisels.
2. Many machine parts are made with high carbon steel since it is capable of withstanding shocks and vibrations.

(*c*) **High tensile steel:** It contains 0.8% carbon and 0.6% manganese. High tensile steel wires are mainly used in prestressed concrete works.

8.5.4 Non-ferrous Materials

The following non-ferrous materials are also used in construction:

(*a*) Copper (*b*) Lead
(*c*) Tin (*d*) Zinc
(*e*) Brass (*f*) Bronze
(*g*) German silver (*h*) Aluminium.

(*a*) **Copper:** It is a naturally available metal in the form of ores, which contain small amount of iron and sulphur. After removing impurities it is processed electrolytically to get purest metal. This metal is almost indestructible. Copper scrap can be reprocessed easily to get original copper.

Properties

1. It is having reddish brown colour.
2. It's structure is crystalline.
3. It is highly ductile and malleable.
4. It resists corrosion.
5. It can be welded at red heat condition.
6. Dents on copper can be hammered out.
7. It has high electrical and thermal conductivity.
8. Its melting point is at 1083°C.
9. It can be easily drawn into wires.

Uses

1. It is used as electric wire and cable.
2. It is used as lightening conductor.
3. For water proofing the construction joints copper plates are useful.
4. Copper tubes are used for hot and cold water supply, gas and sanitation services.
5. It forms a major constituent of brass and bronze.

(*b*) **Lead:** It occurs in the form of lead sulphide which is first converted as lead oxide and then reduced by a carbon to metal.

Properties

1. It is a densest metal, having specific gravity of 11.34.
2. It is soft and highly ductile.
3. It is highly resistant to corrosion and chemical attacks.
4. It is a good conductor of electricity and heat.
5. Its melting point is 326°C.
6. It marks on paper.

Uses

1. Used as sanitation pipes, gutters and cistern linings.
2. Used as covering for cables.
3. It is used for soldering in plumbing works.
4. It is commonly used for ornamental lead works.
5. Lead oxide is useful for paints of different colours.

(*c*) **Tin:** It occurs chiefly as tinstone (SnO_2). This ore is processed to get tin.

Properties

1. It is a silvery white metal.
2. It is soft and malleable.
3. It becomes brittle at 200°C and melts at 230°C.
4. It dissolves in hydrochloric acid.
5. It withstands corrosion due to acid.
6. It is good conductor of heat and electricity.

Uses

1. It is used for giving protective coating to copper and iron utensils.

2. It is used in preparation of solders.
3. Tin foils are used for silvering of mirrors, wrapping of cheese, chocolates etc.

(*d*) **Zinc:** In nature, it occurs in the form of ores of red zinc ZnO, franklinite ZnC, Fe_2O_3, zinc copper $ZnCO_3$ and zinc blende ZnS. Zinc blende is the chief source of the metal.

Properties

The important properties of zinc are as follows:

1. It is a bluish white crystalline metal.
2. It becomes ductile between 100 and 150°C and hence it can be drawn into wire or sheet at this temperature.
3. It melts at 420°C.
4. Its specific gravity is 7.0.
5. It is a good conductor of heat and electricity.
6. It is harmfully attacked by acid, hot and marine water.

Uses

1. It is used to provide protective coating to iron sheets and pipes
2. It is used for preparing paints and batteries.
3. It is useful for making alloys like brass and german silver.

(*e*) **Brass:** It is an alloy of copper and zinc. Its resistance to corrosion is high. It is used for making household utensils. It is used in certain machine parts.

(*f*) **Bronze:** It is an alloy of copper and tin. It consists of 90 per cent of copper. Its resistance to corrosion is high. It can take fine polish. It is used for making nuts, bolts and bearings. It is used in many parts of ship machineries, since its resistance to marine atmosphere is good.

(*g*) **German silver:** It is an alloy of 20 to 30 per cent zinc, 45–60% copper and nickel. It is having high resistance to corrosion. It is used for making resistance coils, for making keys of type writers and musical instruments.

(*h*) **Aluminium:** It is present on the surface of earth crust in most of rocks and clay. But to produce aluminium bauxite Al_2O_3, $2H_2O$ is ideally suited ore.

Properties

1. It has silver colour and bright lustre.
2. It is very light in weight.
3. It is a good conductor of electricity.
4. It is highly resistant to corrosion.
5. It melts at 660°C.
6. It is highly ductile and malleable.
7. It has high strength weight ratio.

Uses

1. It is used to make door and window frames.
2. Aluminium structural members are becoming popular.
3. Aluminium wires, bars and rods, etc. are used for various purposes.
4. It is used as a foil.
5. It's powder servers as pigments in paints.

8.6 CERAMIC MATERIALS

Ceramics are the clay products made by potter's art. The following are the important ceramics.

(*a*) Tiles
(*b*) Terracotta
(*c*) Glazed Earthenwares tiles
(*d*) Stoneware
(*e*) Porcelain

(*a*) **Tiles:** These are resembling bricks but are thin. Depending upon their use, building tiles may be further classified as:

(*i*) roofing tiles
(*ii*) flooring tiles
(*iii*) wall tiles.

Mangalore tiles and Allahabad tiles are popularly used roofing tiles. They are made with stiff clay pressed in the mould and then burnt. Flooring tiles and wall tiles are manufactured by burning pressed green tiles twice. First they are burnt at 700°C. Then they are dipped in the glaze solution and burnt again at 1250°C to fuse them to glaze. The thickness of all these tiles vary from 15 to 20 mm.

IS 654–1972 specifies requirements for mangalore pattern roofing tiles. IS 1478–1969 specifies for flooring tiles. In general the requirements of good tiles are:

1. Tolerance for length ± 5 mm
2. Tolerance for thickness = 2 mm
3. Should be uniform in size and shape.
4. Warpage should not exceed 2% along the edges and 1.5% along the diagonal.
5. They should possess uniforms colour.
6. They should be sound hard and durable.
7. Should give a clear ringing sound when struck with each other.
8. Should have very low percentage of water absorption.

(*b*) **Terracotta:** Terracotta means baked earth. It is used as a substitute to stone for the ornamental works in buildings. It is of brownish-red colour and is used in the form of shaped blocks. Clay and brick mixture is ground finely and mixed with water thoroughly. After draining out water, it is moulded and slowly dried. Then it is burnt at 1100°C to 1200°C.

(*c*) **Glazed earthenware tiles:** Wherever cleanliness is an important criteria, glazed earthen ware tiles are used for floor and wall finishes. They are commonly used in kitchens, bathrooms and hospitals. They are glazed only on exposed side. The colour and texture of tiles should be uniform.

(*d*) **Stoneware:** It is made with refractory clay mixed with crushed pottery and stone. After moulding, the product is burnt for 1 to 3 days. If it is to be used for resisting acids, salt glazing is done. The stoneware is impervious to moisture and is hard. It is used for making sewer pipes, sanitary fitting, gully traps, vessels for holding chemicals, etc.

(*e*) **Porcelain:** It is white and semitransparent. The clay of sufficient purity is used for preparing porcelain. Its constituents are quartz and felspar. They are all mixed in liquid state and moulded to desired shape. Porcelain is used for manufacturing sanitary wares, insulators, storage vessels, crucible, etc.

8.7 TIMBER

Timber refers to wood used for construction works. In fact the word timber is derived from an old English word 'Timbrian' which means 'to build'. A tree that yields good wood for construction is called 'Standing Timber'. After felling a tree, its branches are cut and its stem is roughly converted into pieces of suitable length, so that it can be transported to timber yard. This form of timber is known as 'rough timber'. By sawing, rough timber is converted into various commercial sizes like planks, battens, posts, beams etc. Such form of timber is known as converted timber.

Timber was used as building material even by primitive man. Many ancient temples and palaces and bridges can be seen even today.

Properties of Good Timber:

Colour: It should be uniform.

Odour: It should be pleasant when cut freshly. Unpleasant small indicates decaying timber.

Soundness: A clear ringing sound when struck indicates the timber is good.

Texture: Texture of good timber is fine and even.

Grains: In good timber grains are close.

Density: Higher the density better is the timber.

Hardness: Harder timbers are strong and durable.

Warping: Good timber will not warp under changing environmental conditions.

Toughness: Timber should be capable of resisting shock loads.

Abrasion: Good timber does not deteriorate due to wear. This property is very much required if the timber is used for flooring.

Strength: Timber should have high strength in bending, shear and direct compression.

Modulus of Elasticity: Timber with higher modulus of elasticity is preferred in building construction.

Fire Resistance: A good timber should have high resistance to fire.

Permeability: Good timber has low water permeability.

Workability: Timber should be easily workable. It should not clog the saw.

Durability: Good timber is one which is capable of resisting the actions of fungi, insects, chime attack and changing weather conditions. By seasoning and treating with preservatives durability of timber can be improved.

Defects: Good timber is free from defects like dead knots, shakes and cracks.

Uses of Timber: Timber is used for the following works:

(*a*) For heavy construction works like columns, trusses, piles.

(*b*) For light construction works like doors, windows, flooring and roofing.

(*c*) For other permanent uses like for sleepers, fencing poles, gates and electric poles.

(*d*) For temporary works in construction like scaffolding and centering works, shoring and struthing during excavation of soft soil, packing material.

(*e*) For decorative works like showcases and furniture works.

(*f*) For body works of busses, trains, lorries and boats.

(*g*) For industrial uses like pulps used in manufacturing of papers, card boards, wall papers, etc.

(*h*) For making sports and musical instruments.

8.8 SAND

Sand is a natural product which is obtained as river sand, nalla sand and pit sand. However sea sand should not be used in making mortar and concrete for the following reasons:

1. It contains salt and hence structure will remain damp. The mortar is affected by efflorescence and blisters appear.
2. It contains shells and organic matter, which decompose after some time reducing life of mortar and concrete.

Sand can be obtained artificially by crushing stones. White crushing stones to get coarse aggregates, it is obtained as a by-product. The minute particles form artificial sand for construction works. In constructing dams and bridges artificial sand is commonly used.

Sand is used in mortar and concrete for the following purpose:

1. It subdivides the paste of binding material into thin films and allows it to adhere and spread.
2. It fills up the gap between the building blocks and spreads the binding material.
3. It adds to the density of mortars and concrete.
4. It prevents shrinkage of cementing material.
5. It allows carbon dioxide from the atmosphere to reach some depth and thereby improve setting power.

6. The cost of cementing material per unit volume is reduced as this low cost material increases the volume of cementing material.
7. Silica of sand contributes to formation of silicates resulting into hardened mass.

The properties of good sand

(*a*) It should be chemically inert.
(*b*) It should be free from organic or vegetable matter. Usually 3 to 4% clay is permitted.
(*c*) It should be free from salt.
(*d*) It should contain sharp, angular and coarse grains.
(*e*) It should be well graded.

Uses of Sand

Sand is used as a base coarse to place flooring tiles so as to get level surface. It is mainly used as an inert material for making mortar and concrete.

8.9 AGGREGATE (COARSE)

Sand is known as fine aggregate and broken stones as coarse aggregate which are generally used as aggregates in manufacturing concrete. The broken stone retained on BIS test sieve No. 480 is termed as coarse aggregate. The maximum size of aggregate used depends upon the type of work. For foundations and mass concrete works aggregates of size 38 mm down size are used, while for reinforced concrete works 20 mm down size are preferred.

Properties of Good Aggregates are

1. It should be hard and durable.
2. It should be free from organic impurities.
3. Clay content should be below 4%.
4. It should be well graded.

Uses of Coarse Aggregates

1. Used as an inert material in concrete.
2. Used as railway ballast.
3. Used as base coarse material for road works.

8.10 MORTAR

Mortar is a mixture of

(*a*) Binding material
(*b*) Fine aggregate (sand) and
(*c*) Water.

When water is added to the mixture of binding material and inert material sand, hydration of binding material starts and it binds not only inert material but also the surrounding stones and bricks. If cement is the binding material, then the mortar is called as cement mortar. Other mortars commonly used are lime mortar and mud mortar. River sand is preferred as fine aggregate since it is free from harmful materials like clay.

For preparing mortar, first a mixture of cement and sand is made thoroughly mixing them in dry condition. Water is gradually added and mixed with shovels. The cement to sand proportion recommended for various works is shown in Table 8.2.

Table 8.2 Cement: sand proportion for various works

Sr. No.	*Work*	*Cement : Sand*
1	Masonry Work	1 : 6 to 1 : 8
2	Plastering Masonry	1 : 3 to 1 : 4
3	Plastering Concrete	1 : 3
4	Pointing	1 : 2 to 1 : 3

Properties of Cement Mortar

1. When water is added to a dry mixture of cement and sand, hydration of cement starts and it binds sand particles and the surrounding surfaces of masonry and concrete.
2. The strength of mortar depends upon the proportion of cement and sand. Strengths obtained with various proportion of cement and sand is shown in Table 8.3.

Table 8.3

Sr. No.	*Cement : Sand*	*Compressive Strength*
1	1 : 3	10 N/mm^2
2	1 : 4	7.5 N/mm^2
3	1 : 5	5.0 N/mm^2
4	1 : 6	3.0 N/mm^2
5	1 : 8	0.7 N/mm^2

3. A mix richer than 1 : 3 is prone to shrinkage.
4. Well proportioned mortar provides impervious surface.
5. Leaner mix is not capable of closing the voids in sand and hence the plaster surface is porous.

Uses of Cement Mortar

Mortar is used

1. To bind masonry units like stones, bricks and hollow cement blocks.
2. To plaster slabs and walls to make them impervious.
3. To give neat finishing to concrete works.
4. For pointing masonry joints.
5. For preparing hollow blocks.
6. As a filler material in ferro cement works.
7. To fill joints and cracks in walls.
8. As a filler material in stone masonry.

8.11 PLAIN CONCRETE

The major ingredients of concrete are

1. Binding material (like cement, lime, polymer)
2. Fine aggregate (sand)
3. Coarse aggregates (Crushed stones, jelly)
4. Water

A small quantity of admixtures like air entraining agents, water proofing agents, workability agents, etc. may also be added to impart special properties to the plain concrete mixtures.

When water is added to the dry mixture of binding materials and the aggregates, it forms a plastic mass which can be easily moulded to the desired shape and size. After a stipulated time, the moulded mass hardens and becomes solid mass. Depending upon the binding materials used, there are three types of concrete:

1. Lime concrete
2. Cement concrete
3. Polymer concrete

Lime concrete and polymer concrete have limited uses, whereas cement concrete is predominantly used for construction.

Properties of Concrete

Concrete has completely different properties when it is in the plastic stage and when hardened. Concrete in the plastic stage is also known as green concrete. The properties of green concrete include

(*a*) Workability
(*b*) Segregation
(*c*) Bleeding and
(*d*) Harshness

The properties of the hardened concrete are

(*a*) Strength
(*b*) Resistance to wear and tear
(*c*) Dimensional changes
(*d*) Durability
(*e*) Impermeability

Properties of Green Concrete

Workability: This is defined as the ease with which concrete can be compacted fully without segregation and bleeding. It can also be defined as the amount of internal work required to fully compact the concrete to optimum density. The workability depends upon the quantity of water, grading, shape and percentage of the aggregates present in the concrete.

Workability is measured by

1. The slump observed when the frustum of the standard cone filled with concrete is lifted and removed.
2. The compaction factor determined after allowing the concrete to fall through the compaction testing machine.
3. The time taken in seconds for the shape of the concrete to change from the slump cone to cylinder when tested in the Vee-Bee Consistometer.

The suggested values of workability for different applications are as shown in Table 8.4.

Table 8.4 Suggested mean value of workability

Application	*Slump*	*Compaction factor*	*Time in Vee-Bee Consistometer*
Concreting of shallow sections with vibrators	–	0.75–0.80	10–20
Concreting of lightly reinforced sections with vibrators	–	0.80–0.85	5–10
Concreting of lightly reinforced sections without vibration and heavily reinforced sections with vibration	25–75 mm for 20 mm aggregates	0.85–0.92	2–5
Concreting of heavily reinforced sections without vibration	75–125 mm for 20 mm aggregates	More than 0.92	–

***Segregation*:** Separation of coarser particles from the green concrete is called segregation. This may happen due to lack of sufficient quantity of finer particles in concrete or due to throwing of the concrete from greater heights at the time of placing the concrete. Because of segregation, the cohesiveness of the concrete is lost and honeycombing results. Ultimately it results in loss of strength of hardened concrete. Hence utmost care is to be taken to avoid segregation.

***Bleeding*:** This refers to the appearance of water along with cement particles on the surface of the freshly laid concrete. This happens when there is excessive quantity of water in the mix or due to excessive compaction. Bleeding causes the formation of pores and renders the concrete weak. Bleeding can be avoided by suitably controlling the quantity of water in the concrete and by using finer grading of aggegates.

***Harshness*:** Harshness is the resistance offered by concrete to its surface finish. Harshness is due to the presence of lesser quantity of finer aggregates, lesser cement mortar and the use of poorly graded aggregates. It can also result due to insufficient quantity of water. With harsh concrete, it is difficult to get a smooth surface finish and the concrete would be porous.

Properties of Hardened Concrete

***Strength*:** The characteristic strength of concrete is defined as the compressive strength of 150 mm size cubes after 28 days of curing below which not more than 5 per cent of the test results are expected to fail. The unit of measurement is N/mm^2. IS 456 grades the concrete based on its characteristic strength as shown in Table 8.5.

Table 8.5 Grades of concrete

Grade designation	*Characteristic strength*
M_{10}	10 N/mm^2
M_{15}	15 N/mm^2
M_{20}	20 N/mm^2
M_{25}	25 N/mm^2
M_{30}	30 N/mm^2
M_{35}	35 N/mm^2
M_{40}	40 N/mm^2

Till recently M_{15} concrete was permitted to be used for reinforced concrete work. But IS 456–2000 specifies minimum grade of M_{20} to be used for reinforced concrete works.

Strength of concrete depends upon amount of cement content, quality and grading of aggregates, water cement ratio, compaction and curing. Strength of concrete increases very fast in the initial stages. In 7 days the strength gained is as much as 60 to 65 per cent of 28 days strength. It is customary to assume the 28 days strength as the full strength of concrete. However concrete gains strength beyond 28 days also. The characteristic strength may be increased by the age factor given in Table 8.6.

Table 8.6 Age factor based on expected design loads

Minimum age of member when design load is expected	*Age factor*
1 month	1.0
3 months	1.10
6 months	1.15
12 months	1.20

The tensile strength may be estimated from the formula

$$f_t = 0.7\sqrt{f_{ck}} \text{ N/mm}^2$$

where f_{ck} is the characteristic compressive stress. The modulus of elasticity may be estimated from the formula

$$E = 500\sqrt{f_{ck}} \text{ N/mm}^2$$

***Dimensional change*:** Concrete shrinks with age. The total shrinkage depends on the constituents of concrete, size of the member and the environmental conditions. Total shrinkage is approximately 0.0003 of original dimension.

The permanent dimensional changes due to loading is termed as creep. Its value depends upon the stress in concrete, the age of the concrete at the time of loading and the duration of loading. The ultimate creep strain may be estimated from the values of creep coefficient. The creep coefficient is defined as ultimate creep strain/elastic strain at the age of loading. These values are listed in Table 8.7.

Table 8.7 Creep coefficient based on age of loading

Age of loading	*Creep coefficient*
7 days	2.2
28 days	1.6
1 year	1.1

The size of concrete may change due to thermal expansion also. The coefficient of thermal expansion depends on the nature of cement, the aggregate, the cement content, the relative humidity and the size of sections of the structural elements. Table 8.8 gives the coefficient of thermal expansion for concrete with different types of aggregates.

Table 8.8 Coefficient of thermal expansion

Type of aggregate	*Coefficient of thermal expansion per degree centigrade*
Quartzite	$(1.2 \text{ to } 1.3) \times 10^{-5}$
Sand stone	$(0.9 \text{ to } 1.2) \times 10^{-5}$
Granite	$(0.7 \text{ to } 0.95) \times 10^{-5}$
Basalt	$(0.8 \text{ to } 0.95) \times 10^{-5}$

***Durability*:** Durability refers to the resistance to forces of environment such as weathering, chemical attack and fire. Freezing and thawing result in adverse stress in cold climates. Generally, dense and strong concretes have better durability. The cube crushing strength alone is not a reliable guide to the quantity and durability of concrete. It should have an adequate cement content and should have a low water cement ratio.

Impermeability**:** This is the resistance of concrete to the flow of water through the pores. Excess water during concreting leaves a large number of continuous pores leading to permeability. Since the permeability reduces the durability of concrete, it should be kept very low by using low water cement ratio, dense and well graded aggregates, proper compaction and continuous curing at low temperature conditions. The cement content used should be sufficient to provide adequate workability with low water cement ratio and the available compaction method.

Desirable Properties of Concrete

Appropriate quality and quantities of cement, fine aggregate, coarse aggregate and water should be used so that the green concrete has the following properties:

(*a*) Appropriate workability which depends on the section of concrete element and compaction techniques used.

(*b*) Segregation should not take place while transporting and placing.

(*c*) Bleeding should not take place after placing.

(*d*) Concrete surface should not be harsh.

Hardened concrete should have

(*a*) Developed the required characteristic strength for which structural element is designed.

(*b*) Minimum dimensional changes.

(*c*) Good durability.

(*d*) Good permeability (permit as little water as possible to flow through it).

(*e*) Good resistance to wear and tear, if used for flooring and road work.

Uses of Concrete

1. Bed concrete below the column footings, wall footings and beams.
2. Sill concrete to get a hard and even surface at sill.
3. Over the parapet walls as coping concrete.
4. For flagging the area around buildings.
5. For making tennis courts, basket ball courts.
6. For pavements.
7. To make concrete building blocks as an alternate to bricks.

However major use of concrete is as part of reinforced or pre-stressed concrete. Many structural elements like footings, columns, beams chajjas, lintels, roofs are made with reinforced cement concrete. Cement concrete is used for making storage structures like water tanks, bins, silos, bunkers, etc. Bridges, dams, retaining walls are R.C.C. structures in which concrete is the major ingredient.

Reinforced Cement Concrete (R.C.C.)

Concrete is good in resisting compressive stress but is very weak in resisting tensile stress. Hence reinforcement is provided in the concrete wherever tensile stress is expected. The best reinforcement is steel, since the tensile strength of steel is quite high and the bond between steel and concrete is good. As the elastic modulus of steel is high, the force developed in steel is high. However in tensile zone, hair cracks in concrete are unavoidable. Reinforcements are usually in the form of mild steel or ribbed steel bars of diameter 6 to 32 mm. A cage of reinforcements is prepared as per the design requirements, kept in a form work and then green concrete is poured. After the concrete hardens, the form work is removed. The composite material of steel and concrete, now called R.C.C. acts as a structural member and can resist tensile as well as compressive stress efficiently.

Properties of R.C.C.

1. It should be capable of resisting tensile, compressive, bending and shear stresses.
2. It should not show excessive deflections and spoil the serviceability requirement.
3. There should be proper cover to reinforcement so that corrosion of steel is prevented.
4. The hair cracks developed should be within permissible limit.
5. R.C.C. is a good fire resistant.
6. It can be moulded to desired shape and size.
7. Durability of R.C.C. is very good.
8. R.C.C. structure can be designed to take any load.

Uses of R.C.C.

1. R.C.C. is used as a structural element, wherever direct tension or bending tension is expected. The common structural elements in a building where R.C.C. is used are
   (*a*) Footing (*b*) Columns
   (*c*) Beams, Lintels (*d*) Chejjas, roof slabs, curved roofs, etc.
   (*e*) Stairs
2. R.C.C. is used for the construction of storage structures like
   (*a*) Water tanks (*b*) Dams
   (*c*) Bins (*d*) Silos and Bunkers
3. They are used for the construction of big structures like
   (*a*) Bridges (*b*) Retaining walls
   (*c*) Docks and harbors (*d*) Under water structures
4. R.C.C. is used for pre-casting
   (*a*) Railway sleepers (*b*) Electric poles
5. R.C.C. is used for tall structures like
   (*a*) Multi-storey buildings (*b*) Chimneys
   (*c*) Towers
6. R.C.C. is used for paving
   (*a*) Highways (*b*) City roads
   (*c*) Airports
7. R.C.C. is used in atomic plants to prevent danger of radiation. For this purpose, R.C.C. walls built are 1.5 m to 2.0 m thick.

8.12 PAINTS AND VARNISHES

Paints are applied on the surfaces of timber, metals and plastered surfaces as a protective layer and at the same time to get pleasant appearance. Paints are applied in liquid form and after some time the volatile constituent evaporates and hardened coating acts as protective layer. The varnishes are transparent solutions of resinous materials and they are applied over the painted surfaces.

8.12.1 Paints

The essential constituents of paints are:

(*a*) Base (*b*) A vehicle

(*c*) A pigment (*d*) A drier and

(*e*) A thinner.

(*a*) ***Bases:*** It is a principal constituent of paint. It also possesses the binding properties. It forms an opaque coating. Commonly used bases for paints are white lead, red lead, zinc oxide, oxide of iron, titanium white, aluminium powder and lithophone. A lead paint is quite suitable for painting iron and steel work as it sticks to them quite well. However it is affected by atmospheric action, hence it is not to be used as final coat. White zinc forms good base but is costly. Lithophone which is a mixture of zinc sulphide and barytes is cheap, gives good appearance but is affected by day light. Hence this is used for interior works.

(*b*) ***Vehicles:*** The vehicles are the liquid substances which hold the ingredients of a point in liquid suspension and allow them to be applied on the surface to be painted. Linseed oil, Tung oil, Poppy oil and Nut oil are used as vehicles in paint. Of the above four oils linseed oil is very commonly used vehicle. Boiling makes the oil thicker and darker. Linseed oil reacts with oxygen and hardens by forming a thin film.

(*c*) ***Pigment:*** Pigments give required colour for paints. They are fine particles and have a reinforcing effect on thin film of the paint. The common pigments for different colours are:

Black—lamp black, suit and charcoal black.

Red—venedian red, red lead and Indian red.

Brown—burned timber, raw and burned sienna.

Green—chrome green, copper sulphate.

Blue—prussian blue and ultra marine.

Yellow—ochre and chrome yellow.

(*d*) ***The Drier:*** These are the compounds of metals like lead, manganese, cobalt, etc. The function of a drier is to absorb oxygen from the air and supply it to the vehicle for hardening. The drier should not be added until the paint is about to be used. The excess drier is harmful because it destroys elasticity and causes flaking.

(*e*) ***The Thinner:*** It is also known as solvent. It makes paint thinner and hence increases the coverage. It helps in spreading paint uniformly over the surface. Terpentine and neptha are commonly used thinners. After paint is applied thinner evaporates and paint dries.

Properties of an Ideal Paint

1. It should be possible to apply easily and freely.
2. It should dry in reasonable time.
3. It should form hard and durable surface.
4. It should not be harmful to the health of workers.
5. It should not be easily affected by atmosphere. It should maintain its colour for longer time.
6. It should possess attractive and pleasing appearance.
7. It should form a thin film of uniform nature, *i.e.*, it should not crack.
8. It should possess a good spreading power.
9. It should be cheap and economical.

Types of Paints: Depending upon their constituents there are various paints. A brief description of some of them, which are commonly used, are given below:

(*a*) ***Oil Paints:*** These are applied in three coats—primer, undercoat and finishing coat. This is cheap and easy to apply. The presence of dampness while applying adversely affects the life of oil paints.

(*b*) ***Enamel Paint:*** It contains white lead, oil, petroleum spirit and resinous material. The surface provided by it resists acids, alkalies, and water very well. It is desirable to apply a coat of titanium white before the coat of enamel is applied. It can be used both for external and internal walls.

(*c*) ***Emulsion Paint:*** It contains binding materials such as polyvinyl acetate, synthetic resins, etc. It dries in $1\frac{1}{2}$ to 2 hours and it is easy to apply. It is more durable and can be cleaned with water. For plastered surfaces first a coat of cement paint should be applied and then the emulsion paint. Emulsion paint needs sound surface.

(*d*) ***Cement Paint:*** It is available in powder form. It consists of white cement, pigment and other additives. It is durable and exhibits excellent decorative appearance. It should be applied on rough surfaces rather than on smooth surfaces. It is applied in two coats. First coat is applied on wet surface, but free from excess water and allowed to dry for 24 hours. The second coat is applied within 24 hours which gives good appearance.

(*e*) ***Water Paints:*** These paints are also known as distempers. They contains water as thinning agent. These paints are suitable only for interior walls.

(*f*) ***Bituminous Paints:*** This paint is manufactured by dissolving asphalt or vegetable bitumen in oil or petroleum. It has black colour. It is used for painting iron works under water.

(*g*) ***Synthetic Rubber Paint:*** This paint is prepared from resins. It dries quickly and is little affected by weather and sun light. It resists chemical attack well. This paint may be applied even on fresh concrete. Its cost is moderate and it can be applied easily.

(*h*) ***Aluminium Paint:*** It contains finely ground aluminium in spirit or oil varnish. It is visible in darkness also. The surfaces of iron and steel are protected well with this paint. It is widely used for painting gas tanks, water pipes, oil tanks, etc.

(*i*) ***Anti-Corrosive Paint:*** It consists essentially of oil, a strong drier, lead or zinc chrome and finely ground sand. It is cheap and resists corrosion well. It is black in colour.

Application of Paint: Preparation of surface for application of paint is most important part in painting. The surface to be painted should not be oily and it should be free from flakes of the old paint. Cracks in the surface should be filled with putty and then with sand paper, then primer is applied. Painting work should be carried out in dry weather. The under coats and first coats must be allowed to dry before final coat is applied.

8.12.2 Varnishes

Varnish is the solution of resins or resinous substances prepared in alcohol, oil or turpentine. It is applied to timber surfaces. It brightens the appearance of grains in wood and provides a good protection layer. It renders the wooden surface glossy. Depending upon the solvent used varnished are classified as oil varnishes, spirit varnishes, turpentine varnishes and water varnishes.

8.13 GLASS

Silica is the main constituent of glass. But it is to be added with sodium potassium carbonate to bring down melting point and lime or lead oxide to make it durable. Manganese oxide is added to nullify the adverse effects of unwanted iron present in the impure silica. The raw materials are ground and sieved. They are mixed in specific proportion and melted in furnace. Then glass items are manufactured by blowing, flat drawing, rolling or pressing.

Important Properties of Glasses are:

1. It absorbs, refracts or transmits light. It can be made transparent or translucent.
2. It can take up high polish.
3. It is an excellent electrical insulator.
4. It is strong and brittle.
5. It can be blown, drawn or pressed.
6. It is not affected by atmosphere.
7. It is good resistant to chemicals.
8. It is available in beautiful colours.
9. With the advancement in technology, it is possible to make glass lighter than cork or stronger than steel.
10. Glass panes can be cleaned easily.

Types of Glass

The glasses may be broadly classified as

(*a*) Soda-lime glass

(*b*) Potash-lime glass

(*c*) Potash-lead glass and

(*d*) Common glass

(*e*) Special glasses

The properties and uses of the above glasses are briefly presented below:

(*a*) ***Soda Lime Glass:*** It is mainly a mixture of sodium silicate and calcium silicate. It is fusible at low temperature. At low temperature it can be blown or welded. It is colourless. It is used as window glass. Laboratory tubes and apparatus are manufactured with this glass.

(*b*) ***Potash Lime Glass:*** It is mainly a mixture of potassium silicate and calcium silicate. It is also known as hard glass. It fuses at high temperature. It is used in the manufacture of glass articles which have to with stand high temperatures.

(*c*) ***Potash Lead Glass:*** It is mainly a mixture of potassium silicate and lead silicate. It possesses bright lustre and great refractive power. It is used in the manufacture of artificial gems, electrical bulbs, lenses, prisms, etc.

(*d*) ***Common Glass:*** It is mainly a mixture of sodium silicate, calcium silicate and iron silicate. It is brown, green or yellow in colour. It is mainly used in the manufacture of medicine bottles.

(*e*) ***Special Glasses:*** Properties of glass can be suitably altered by changing basic ingredients and adding a few more ingredients. It has now emerged as a versatile material to meet any requirement in engineering. The following is the list of some of the special glasses:

1. Fibre glass
2. Foam glass
3. Bullet proof glass
4. Structural glass
5. Glass block
6. Wired glass
7. Ultraviolet ray glass
8. Perforated glass.

8.14 PLASTIC

Plastic is an organic material prepared out of resin. It may or may not contain fillers, plasticisers, solvents. Plastic may be defined as a natural or synthetic organic material which have the property of being plastic at some stage of their manufacture when they can be moulded to required shape.

Shellac and bitumen are the natural resins used as plastic for a long time. In 1907, backland produced synthetic resin from the reaction of phenol and formaldehyde. This resin is hardened under pressure and heat to produce useful plastic articles.

Types of Plastics

Primarily there are two types of plastics:

(*i*) Thermosetting and

(*ii*) Thermoplastic.

(*i*) ***Thermosetting Plastics:*** It needs momentary heated condition and great pressure during shaping. When heated between the molecules cross linkage is established and chemical reactions take place. During this stage shape can be changed with pressure. This change is not reversible. The scrap of a such plastic is not reusable. Bakelite is an example of such plastic.

(*ii*) ***Thermoplastic:*** In this variety linkage between the molecules is very loose. They can be softened by heat repeatedly. This property helps for repeated use of waste plastic. These plastic need time to cool down and harden. These plastics are to be kept in mould till cooling takes place completely. Bitumen, cellulose and shellac are the examples of this variety of plastics.

Properties of Plastics

The following are the general properties of plastics:

1. **Colour:** Some plastics are completely transparent. Using pigments plastics of any attractive colours can be produced.
2. **Dimensional stability:** It is dimensionally stable to a great extent.
3. **Durability:** Plastic offers great resistance to moisture and chemicals, and hence more durable.
4. **Electrical insulation:** The plastics possess excellent electric insulating property.
5. **Fire-resistance:** The phenol-formaldehyde and urea-formaldehyde plastics resist fire to a great extent and hence they are used as fire proofing materials.
6. **Strength:** The plastics are reasonably strong. Their strength may be increased by reinforcing with various fibrous materials. Attempts are being made to produce structurally sound plastics.
7. **Specific gravity:** The specific gravity of plastics is very low and hence convenient to handle.
8. **Ductility:** The plastics are not ductile and hence fail without giving warning.
9. **Fixing:** Plastics can be bolted, drilled, glued, clamped or simply push fitted in position.
10. **Maintenance:** There is no maintenance cost for plastic articles They do not need painting, polishing, etc.

Uses of Plastics

There are variety of plastics made to suit different uses. The typical uses of plastics in buildings is listed below:

1. Corrugated and plain sheets for roofing.
2. For making jointless flooring.
3. Flooring tiles.
4. Over head water tanks.
5. Bath and sink units.
6. Cistern ball floats.
7. Decorative laminates and mouldings
8. Window frames and doors for bathrooms.
9. Lighting fixtures.
10. Electrical conduits
11. Electrical insulators
12. Pipes to carry cold waters.

8.15 CONDUCTING AND MAGNETIC MATERIALS

There are materials which exhibit change in their non-mechanical properties like electrical or magnetic properties due to changes in the mechanical simulates like stress and strain. The vice versa is also true. These properties are made use to sense changes and activate mechanism to impart necessary resistance. Hence such materials are also known as smart materials. Shape memory alloys, magnetostrictive material, piezoelectric material, electrostrictive materials and electro-rheological fluids are currently available smart materials.

Shape Memory Alloy (SMA): The metallic materials which demonstrate the ability to return to some previously defined shape or size when subjected to appropriate thermal changes. There are wide variety of alloys which exhibit shape memory.

However for commercial exploitation only Niti (Nickel alloys) and copper base alloys such as CuZnAl (Copper zinc aluminium) alloys are found useful. Niti alloys have greater shape memory (up to 8%) compared to copper base alloys (4.5%).

Magnetostrictive Materials: These materials under go deformation when subjected to magnetic field. Most ferro magnetic materials exhibit some measurable magnetostrictive property. The highest known magnetostriction are those of iron alloys containing rare earth elements – Dysposium ($DeFe_2$) or Terrbium ($TbFe_2$).

Piezoelectrical Materials: Piezoelectrical materials are most widely used smart materials. These materials under go surface elongation when an electrical field is applied across them and produce voltage when surface strain is introduced. Figures 8.3 and 8.4 show these phenomenon. Quartz, Rochelle salt and polyvinylidene fluid are commonly used smart materials.

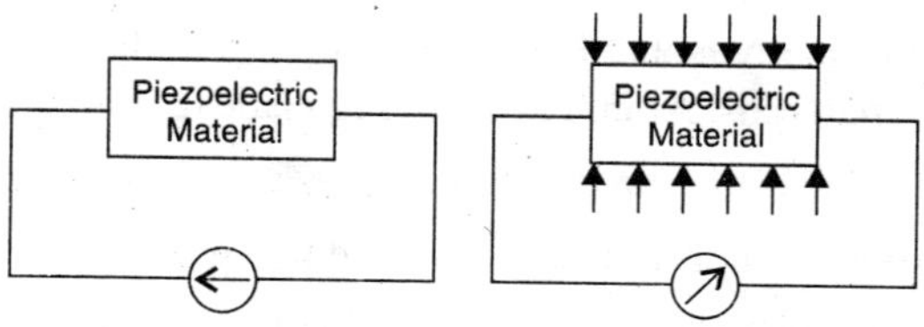

Fig. 8.3 Generation of electricity on application of pressure

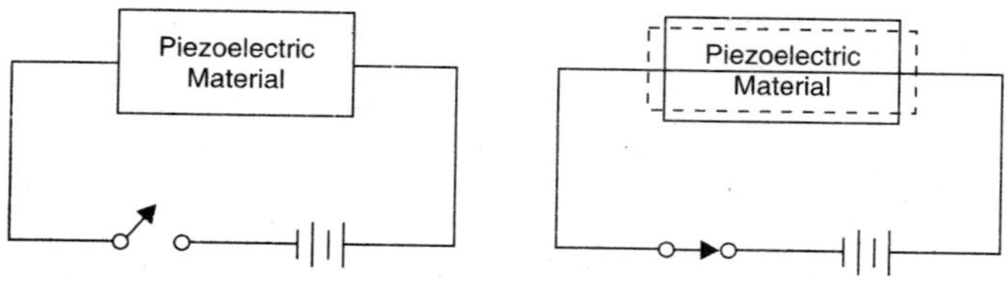

Fig. 8.4 Change in the dimensions of material on application of electricity

Electrostrictive Materials: These are identical to piezoelectrical materials, with better strain capacity but very sensitive to the temperature.

Electro-rheological Fluids: These are colloidal suspensions that exhibit reversible change in viscosity when subjected to an electric field. The property of these materials of reversible transition from liquid to solid state has been utilised in many engineering applications. These changes could be reversed in time intervals of the order of milliseconds.

Applications

In Aircrafts and Spacecrafts: Aircrafts and spacecrafts are large structures which have to float in air. Hence they should be as light as possible. When weight reduction is made from the consideration of only primary forces, the problem of structural instability, excessive deflection and excessive vibration results.

The sensors and actuators are used to dampen the vibration and to reduce deflections. They are fixed at suitable positions to control vibrations. The instabilities are sensed using sensors made of smart materials. Signals are processed in computers and corrective forces are developed through actuators made of smart materials. Figure 8.5 shows the method of building a structure (plate or a beam) so that they can sense and react to lateral instability.

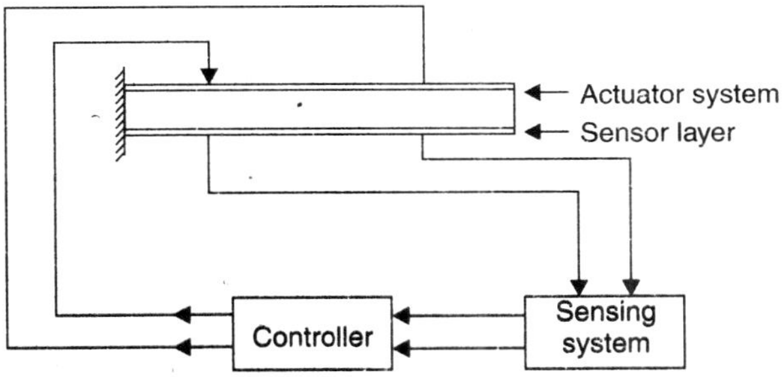

Fig. 8.5 Sensors and actuators

Electro-rheological fluids are used in clutches, valves and engines.

Smart materials can be used to develop safety, security and emergency control systems.

8.16 MISCELLANEOUS MATERIALS

The following are some of the materials which have lot of engineering applications:

1. Bituminous materials
2. Asbestos and
3. Gypsum

8.16.1 Bituminous Materials

Asphalt, bitumen and tar are referred as bituminous materials, which are essentially hydrocarbon materials. The **asphalt** is a mixture of inert mineral matter like alumina, lime, silica, etc. and a hydrocarbon known as asphaltic bitumen. In some places like Trinidad and Bermudez asphalt is available in nature at a depth of 3 to 60 metres. It is known as **natural asphalt.** Common variety used all over the world is **residual asplalt** which is obtained by the fractional distillation of crude petroleum oil. Bitumen is the binding material which is present in asphalt. It is a hydrocarbon. It is obtained by partial distillation of crude oil. It contains 87 per cent carbon, 11 per cent hydrogen and 2 per cent oxygen.

Tar is obtained in the destructive distillation of coal, wood or other organic material. When coal or wood are heated to redness in an closed chamber it yields volatile product and residue coke. They are separated and upon cooling volatile product gives tar.

Comparison between asphalt, bitumen and tar is presented in Table 8.9 below:

Table 8.9. Comparison between asphalt, bitumen and tar

No.	*Property*	*Asphalt*	*Bitumen*	*Tar*
1.	Colour	Blackish brown	Dark with slight reddish tinge	Deep dark
2.	Carbon content	Low	Moderate	High
3.	State	Solid or semisolid	Solid	Viscous liquid
4.	Effect on heating	Burns with a smoke flame and becomes plastic	Melts	Becomes more fluid
5.	Setting time	Less	Less	More
6.	Adhesive power	Less	More	Most
7.	Resistance to acid	More	More	Less
8.	Use	As damp proof coarse, for paints, as roofing felt and for road work	As damp proof coarse and as roofing felt	For preserving timber

8.16.2 Asbestos

Asbestos is a general name for several varieties of fibrous minerals which are available in nature. But presently most of the commercial asbestos produced is 'chrisotile' [$Mg_6Si_4O_{11}(OH)_6 \cdot H_2O$].

Properties of Asbestos:

1. It is flexible, soft and non-porous.
2. It is fire proof and acid proof.
3. It is a good insulator for heat and electricity.
4. When it is mixed with cement and water, it retains shape firmly.
5. Its colour is brown or grey.
6. It can be cut into pieces or can be drilled.
7. It possesses high tensile strength in the direction of its fibres.
8. Its specific gravity is 3.10.

Uses

1. Asbestos cement sheets are the cheapest roofing materials.
2. Asbestos cement pipes are used as downtake pipes in buildings.
3. With bitumen it forms good damp proof layer.

4. It is used for preparing fire proof ropes and clothes.
5. It is used as covering material for fuse and switch boxes.
6. It is useful for insulating boilers, furnaces, etc.

8.16.3 Gypsum

It is hydrated sulphate of calcium. Its chemical composition is $CaSO_4 \cdot 2H_2O$. The gypsum containing up to 70 per cent of $CaSO_4$, $2H_2O$ can be used as a building material. When finely ground gypsum is heated at temperature of 160°C to 170°C to drive out 14.7 per cent of water, plaster of paris is formed. Plaster of paris hardens in 3 to 4 minutes. Retarders like clay, citric acid, gum, starch or sugar may be added to delay the hardening of plaster of paris.

Properties of Gypsum

1. It is light in weight.
2. It is not affected by bacteria.
3. It sets with little change in volume.
4. It has good adhesion to fibrous materials.
5. It is fire resistant.
6. It acts as retarder to setting of cement.

Uses of Gypsum

1. It is used as retarder to setting of cement by cement manufacturers.
2. Gypsum boards are used as substitute to plywood.
3. Plaster of paris is used in artwork for buildings.

QUESTIONS

1. What are the requirements of good building stones ?
2. Explain the terms: Igneous rocks, sedimentary rocks and metamorphic rocks.
3. List various properties of stones.
4. What are the requirements and properties of good building bricks.
5. Write short note on any four types of bricks.
6. List various uses of bricks.
7. Distinguish between fat lime and hydraulic lime.
8. What are the various uses of lime?
9. What are the IS specifications for physical properties of Portland Cement? Explain briefly.
10. Write short notes on any five special types of cement.
11. Distinguish cast iron, wrought iron and steel.

12. List properties of mild steel. How the properties of high carbon steel and high tensile steel differ from that of mild steel?
13. Write short notes on
    (*i*) Brass
    (*ii*) Bronze
    (*iii*) Copper
    (*iv*) Aluminium and
    (*v*) Zinc.
14. What is meant by 'ceramic materials'? Briefly explain : Terracotta, Glazed earthenware and Stoneware.
15. What are the requirements of good tiles?
16. List the properties of good timber.
17. Why sand is used in mortar and concrete?
18. What are the required properties of good sand for construction works ?
19. List various uses of mortar.
20. Briefly explain the properties of a good
    (*i*) Green concrete
    (*ii*) Hardened concrete.
21. What is meant by reinforced cement concrete? Why it has become a popular construction material?
22. Briefly explain the function of essential constituents of paints.
23. What are the requirements of ideal paint?
24. Bring out the differences among oil paints, enamel paints, emulsion paints cement paint and water paints.
25. Distinguish between paints and varnishes.
26. List the important properties of glass.
27. Write short notes on any four types of glasses.
28. Define 'plastic'. Differentiate between Thermosetting and Thermoplastic.
29. List various uses of plastics.
30. What are smart materials? Where are they used ?
31. Bring out the differences among asphalt, bitumen and tar.
32. Write short notes on
    (*i*) Asbestos
    (*ii*) Gypsum.

9

Elements of Building Construction

Every family needs a building to reside. Apart from residential purposes, buildings are required for educational, institutional business, assembly and for industrial purposes. Buildings are required for storage of materials also.

All these buildings should be properly planned, keeping in view the various requirements, making use of good ventilation, lighting and making it aesthetically attractive. Taking suitable loads structural components of buildings like foundation, wall columns, floors, roofs, lintels, stairs and lifts are to be designed. Building services like electric and water supply are to be provided and finishing touch is to be given to building to make it attractive. However in planning and designing the building bye-laws of the statutory authorities should not be violated.

In this chapter elementary principles and requirements residential and industrial buildings are presented. Few layouts are shown. Design loads for different types of buildings is given. Functions of various components and their nominal dimensions are presented. Guidelines are given on building drawings and building bye-laws are introduced.

9.1 ELEMENTARY PRINCIPLES OF PLANNING

Planning of a building depends upon factors like bye-laws of the place, location of site, climatic conditions, surrounding environment, requirements of owner etc. Principle listed below are only the guidelines, they are not hard and fast rules:

1. Aspect
2. Prospect
3. Roominess
4. Furniture requirement
5. Grouping
6. Circulation
7. Elegance
8. Sanitation
9. Flexibility
10. Privacy
11. Practical consideration.

A brief description of each one is given below:

1. **Aspect:** Aspect means arrangement of doors, and windows in the external wall to make good use of nature such as sunshine, breeze, scenery, etc. This term has nothing to do with architectural aspect of outlook of building. Kitchen should have eastern aspect to enjoy morning sunshine, means, kitchen should be located on eastern side of the building to make use of morning sun. The following are the required aspects for various parts of building in the northern hemisphere of earth.

   (*i*) Kitchen—eastern aspect.

   (*ii*) Dining room—southern aspect to enjoy winter sun.

   (*iii*) Drawing and living room—southern or south eastern aspect to enjoy winter sun.

   (*iv*) Bedrooms—western or south western aspect to enjoy breeze in summer.

   (*v*) Reading rooms, classrooms, stairs—northern aspect to enjoy diffused light.

2. **Prospect:** It is about locating and selecting types of doors and windows so as to reveal pleasant features and conceal undesirable features of the building from a person viewing from outside. Projecting windows, providing aesthetically good doors and windows etc. are some of the prospects of planning.

3. **Roominess:** It refers to suitable proportioning of length, width and height of rooms in the building to get maximum benefit from the minimum dimensions. Length to width ratio should be from 1.2 to 1.5. If it is nearly square lot of area is wasted for movement, while it is more than 1.5 it gives the 'tunnel' effect. Doors for rooms should be properly located so that liability, utility and privacy is maximum. Cupboards and lofts should be provided to increase roominess. Proper colours to wall and floor also give roominess effect. Light colour create effect of more space.

4. **Furniture requirement:** In planning residential or office, laboratory, hospital building, positions of required furnitures should be drawn and room dimension, positions of doors, window, ward robes, etc. planned. In case of planning a hostel rooms for two students may need centrally placed door while if it is for three students, it should be near end of a wall. Positions of cots, study tables and cupboards should be drawn and room planned. In designing a living room, positions of sofa, chairs, T.V., etc. should be drawn and size of room and position of doors, availability of circulation area checked. Thus furniture requirement influences the planning of a building to a great extent.

5. **Grouping:** Grouping means disposition of various rooms in the building for the convenience of users. A dining room should be close to the kitchen, while sanitary block should be away from kitchen, but convenient to bedrooms. In case of offices administrative department is located centrally. In factories, various sections are located such that the products move in one direction to get finally assembled after least movement. In residential building grouping is to achieve comfort, privacy and efficiency while in case of other building it is to achieve economical service.

6. **Circulation:** Circulation means space provided for movement from room to room or floor to floor. Passages, lobbies, halls provided serve horizontal circulations in a building while stairs

and lift serve for vertical circulation. Within a room also a portion of it serve for circulation while some other portion serves for utility. The following points should be considered while providing circulation.

(*i*) They should be straight

(*ii*) They should be sufficient

(*iii*) They should be sufficiently lighted and ventilated

(*iv*) Stairs should be easily accessible to all users

(*v*) Sanitary services should have access from every user through passage/lobby.

7. Elegance: Elegance means general effect produced for a viewer from outside. It depends upon proper positioning of doors, windows, ventilations balconies, chajjas, staircases, etc. Elevation should be attractive. The width, height and the projections of a building contribute a lot for the elegance. Taj Mahal is an example, which is famous for its elegance.

8. Sanitation: It includes providing good lighting ventilation and sanitary conveniences in the building.

Light: Natural light provides hygenic atmosphere light should not be glaring but it should be uniformly distributed. Providing windows and ventilations of appropriate size at suitable position contributes a lot for natural lighting. For residential buildings window area to floor area should not be less than $1/10^{th}$ while for school buildings it should be $1/5^{th}$.

Ventilation: It is supply of outside air in the building. It is achieved to mainly by selecting and positioning of doors, windows and ventilations. In case it is not possible to achieve natural ventilation for any part of the building provide ordinary or exhaust fans.

Sanitary convenience include provision of bathrooms, lavatories, urinals, etc. Provision of these are not only necessities but statutory requirement also. Suitable slopes should be given to drain out water. These facilities should be located giving free access to all users.

9. Flexibility: This aspect of planning means a room designed for a specific purpose should be possible to use for other purposes, if there is necessity. A study room may be planned for using as a guest room. If partition is provided between living room and dining room, it is possible to remove partition and use living room plus dining room for the family functions. If independent access is given to backyard from kitchen, backyard can be used for dinner functions. Sometimes a living room may be used for sleeping also if instead of sofa provision is made for sofa cum bed. Thus in planning flexibility also should be considered.

10. Privacy: It is an important principle in planning a building. It should take care of privacy of one room from other room in a building as well as of some parts of the building from neighbouring buildings and from streets. It is ensured by proper grouping of rooms and by suitably providing doors, windows and ventilators. To some extent it may be achieved by screens and curtains. Planning the entrance at appropriate position also contributes a lot in providing privacy.

11. Practical consideration: Economy is one of the major practical consideration in planning. A person may not be in a position to invest more than a certain amount at the time of constructing a building. Based on the amount he can invest, the engineer has to work out the plinth area that can be built. Then the building should be planned, making suitable provision for future expansion. In any case there cannot be any compromise on strength and stability of building. However economy may be achieved by making future renovations in the following:

1. Improving elevations without dismantling any part of building.
2. Improving flooring.
3. Extending building horizontally or vertically without damaging existing building.

Using precast lintels, chejjas, etc. and minimising the number of doors and windows without sacrificing comfort also contribute to economise the construction.

9.2 BASIC REQUIREMENTS OF A BUILDING

The following are the basic requirements of a building:

1. Strength and stability
2. Dimensional stability
3. Resistance to dampness
4. Resistance to fire
5. Heat insulation
6. Sound insulation
7. Protection against termite
8. Durability
9. Security against burglary
10. Lighting and ventilation
11. Comforts and convenience
12. Economics.

1. **Strength and Stability:** Building should be capable of transferring the expected loads in its life safety to the ground. Design of various structural components like slabs, beams walls, columns and footing will ensure safety. None of the structural components should buckle or no portion should overturn.

2. **Dimensional Stability:** Excessive deformation of structural components give a sense of instability and result into cracks in walls, flooring tiles, etc. All components of structure should be designed so that deflections do not exceed permissible values specified in the codes.

3. **Resistance to Dampness:** Dampness of building is a great nuisance and may reduce the life of the building. Proper drainage system for rain water, good plumbing, good concreting and plastering will avoid dampness. Great care should be taken in building constructions to avoid dampness.

4. **Resistance to Fire:** The basic requirement laid down in the codes are:

   (*i*) The structure should not ignite easily.

   (*ii*) In case of fire there should be means to vacate building quickly.

   (*iii*) Building design should be such that spread of fire is slow. This is achieved by locating kitchen so that fire is directed away from building rather than towards the building.

   It may be noted that concrete and masonry (stone or brick) have better resistance to fire while steel and wood have lesser resistance. Hence reduce use of steel and wood in kitchen and in bathrooms with electric heaters.

In public buildings and assembly halls fire alarms should be provided. Staircases should be easily accessible.

5. **Heat Insulation:** A building should be designed so as to insulate it from heat. If it can keep inside cool in summer and warm in winter, there is lot of comfort and saving in electric energy on heat insulation. This can be achieved by

   (*i*) Proper orientation of building

   (*ii*) Using thicker outer walls

   (*iii*) Using hollow bricks or concrete blocks for walls

   (*iv*) Providing weatherproof course on roof.

6. **Sound Insulation:** Buildings should be planned against outdoor and indoor noises. Educational institutions, hospitals and houses close to busy streets need sound insulation. Prolonged exposure to noise pollution causes deafness, nervousness and results into inefficiency.

   Noise pollution can be reduced in the building by suitably planning and constructing the building:

   (*i*) Orienting building suitably so as to keep rooms away from road side.

   (*ii*) Using hollow blocks for walls.

   (*iii*) Plugging doors and window openings tightly.

   (*iv*) Using false ceilings.

   (*v*) By fixing water closet cisterns on outer walls instead of wall common with rooms.

   (*vi*) By fixing water closet pan on a thin pad.

   (*vii*) Pipes passing through walls and floors should be wrapped and held in insulated clips.

7. **Protection from Termite:** Termites are also known as vermins and white ants. Dry wood termites live in dry woods while subterrain termites live inside the ground but live by eating wood. They can enter the building through cracks of 0.5 mm also. A building should be protected from these termites by

   (*i*) Treating the foundations with chemicals at the time of construction.

   (*ii*) Using well seasoned and well treated wood in the buildings.

   Once attack from termites start it is difficult to control. Chemical treatment of building is costly and it eliminates problem for a certain period of 5 to 7 years but not permanently. Hence it is better to go for treatment of foundations at the time of construction of the building itself.

8. **Durability:** Durability of a building is the length of time over which building remains usable. It depends upon the resistance of building materials to the effects of sun, rain, frost, wind etc. Durability is increased by protecting doors and windows with sun shades, chejjas, and providing proper slopes, to drain out rain water. It depends upon good maintenance of the building also on quality of materials used in construction contributes to durability.

9. **Security Against Burglary:** This is a basic need in buildings. By providing thicker walls, using stronger doors and windows in outer walls, improve security against burglary. Providing grills to windows, providing additional shutters to doors improve security. Alarms fitted in walls, roofs also improve security of buildings.

10. **Lighting and Ventilation:** Proper orientation and proper positioning of doors, windows and ventilations improve lighting and ventilation inside a building. For healthy and happy living natural light and ventilations are required. Diffused light should be available inside the building.

11. **Comforts and Convenience:** This is one of the basic requirement of a building. Various units of the building should be properly grouped and integrated keeping in mind comfort and convenience of the users.

12. **Economy:** Economy without sacrificing comfort, convenience and durability is another basic requirement in building design. Circulation area should be minimised. Maintenance cost is to be kept low. All these points should be considered at the time of planning, selecting materials and construction.

9.3 LAYOUT OF RESIDENTIAL BUILDINGS

A residential building essentially consists of the following:

1. Living area
2. Sleeping area
3. Service area

1. **Living Area:** This area is for general use. It is provided near the entrance. The following points are considered in planning it:

   (*i*) It should be specious to accommodate furnitures, showcases and should provide sufficient distance to view T.V.

   (*ii*) It should be adjacent to dining room.

   (*iii*) It should be sufficiently lighted.

   (*iv*) It should have a southern or south eastern aspect.

2. **Sleeping Area:** This area is for sleeping and relaxing. The size should be sufficient to accommodate required number of cots, easy chairs and cupboards. They should be provided with attached bathrooms. Western or south western aspect is preferred so as to enjoy breeze in summer.

3. **Service Area:** Kitchen, dining room, passages and bathrooms and other areas constitute service area.

   (*i*) *Kitchen:* It is to be located adjacent to dining room. A cooking platform of width 750 to 800 mm should be provided. Water taps, sink, provision for keeping gas stove, fridge utensils, etc. should be made in the kitchen. Exhaust fan should provided. It should have eastern or north eastern aspect to get benefit of morning sun.

   (*ii*) *Dining room:* This is to be located between drawing hall and kitchen. Necessary partitions may be provided between drawing and dining room and between kitchen and dining room. It should be sufficiently large to accommodate a dining table with 4 to 6 chairs. A service window may be provided between kitchen and dining room.

   (*iii*) *Bathrooms:* Bath room and water closets should be accessible from all rooms. The size of these rooms are governed by the type of fixtures selected. In both room there should be provision for hot water supply also. The walls and floors should be provided with marbles/glazed tiles or finishing with water proof cement.

(*iv*) *Other rooms:* Office rooms are provided near living room, while study rooms are preferred near bed rooms. Pooja room and store rooms is ideally located near kitchen. Varandha is provided near the main entrance. Passages are provided to give independent access to both rooms. Staircases should be accessible easily. The area for the passage and stairs should be small, so that economy is achieved in building construction.

Layout of residential building should meet needs of the owner and the requirements of minimum standards for various components of buildings. Due consideration should be given to the economy and aesthetic look. As somes of these requirements contradict each other, it is finally the art of satisfying the customer (owner). Different people have different customs, needs and satisfaction level. An architect/ civil engineer has to educate the customer about advantages and disadvantages of different layouts and finally select the layout of customer's satisfaction. Typical layout plans are presented in Figures 9.1 to 9.6.

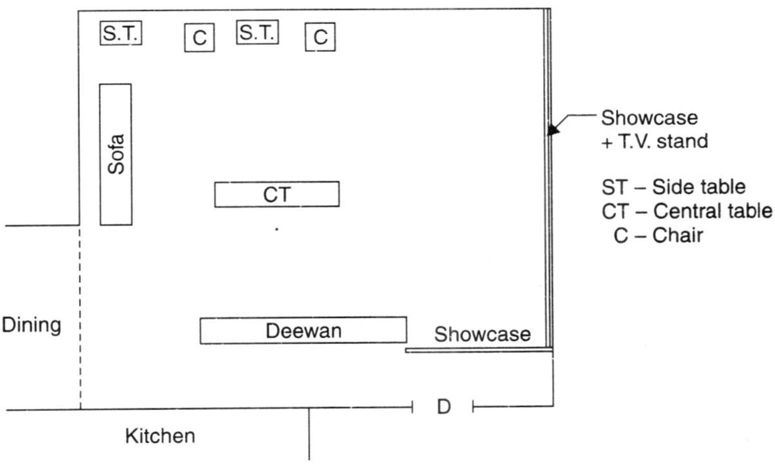

Fig. 9.1 A typical drawing room

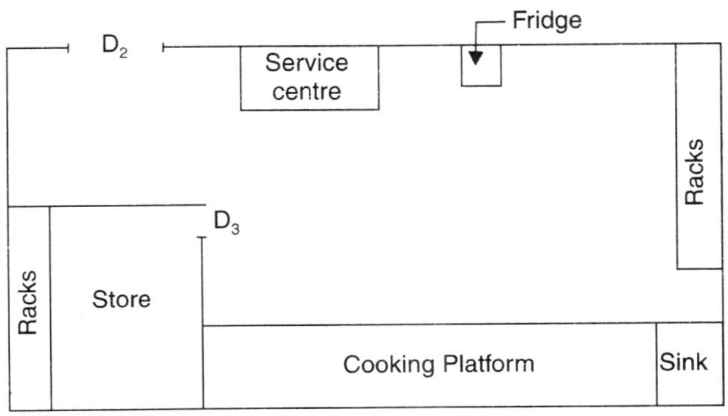

Fig. 9.2 A typical kitchen

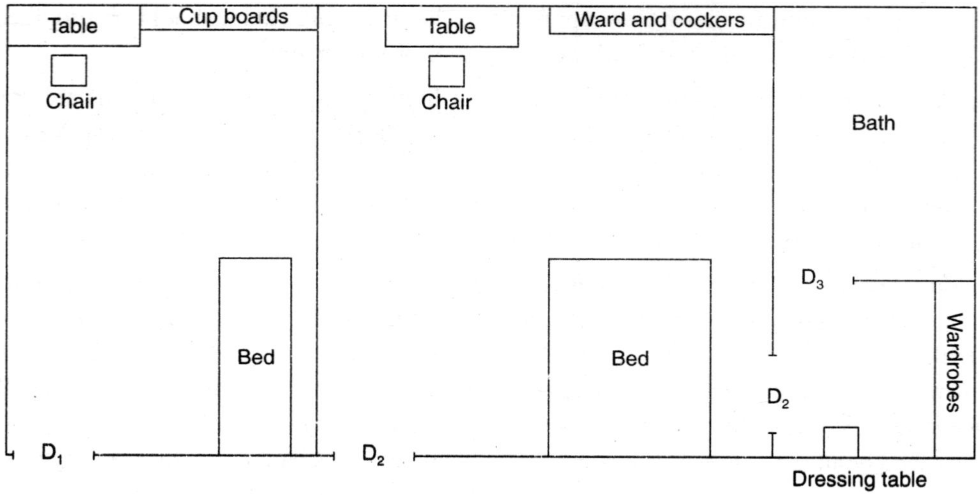

Fig. 9.3 Layout of single and master bedrooms

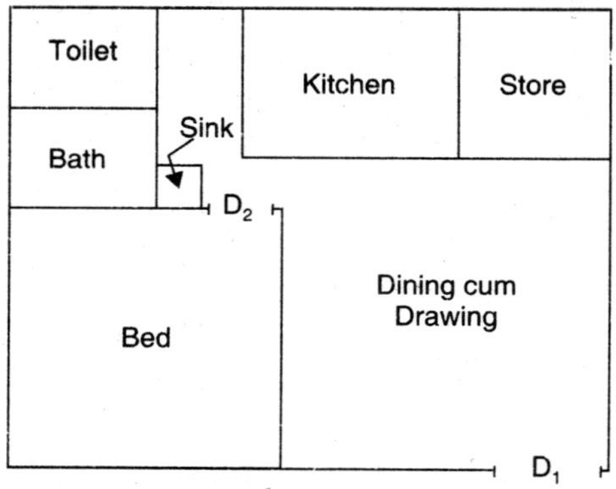

Fig. 9.4 Layout of a single bedroom house

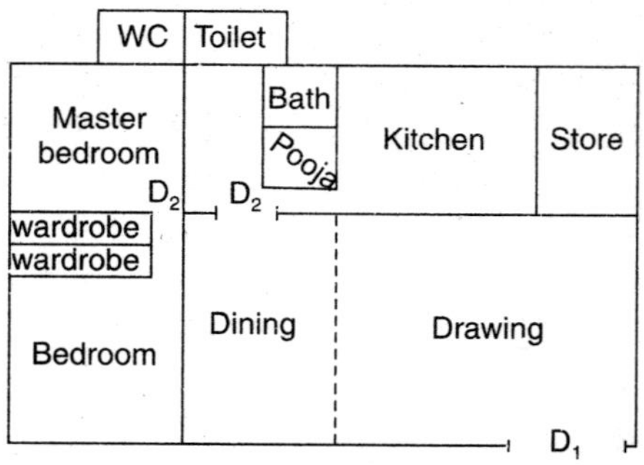

Fig. 9.5 A two bedroom house

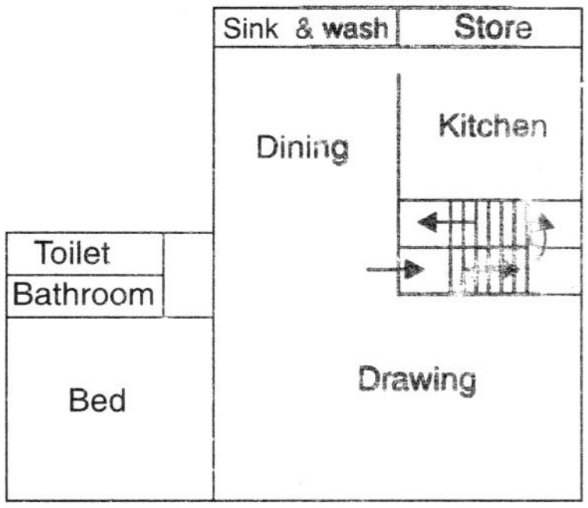

(a) Ground floor plan

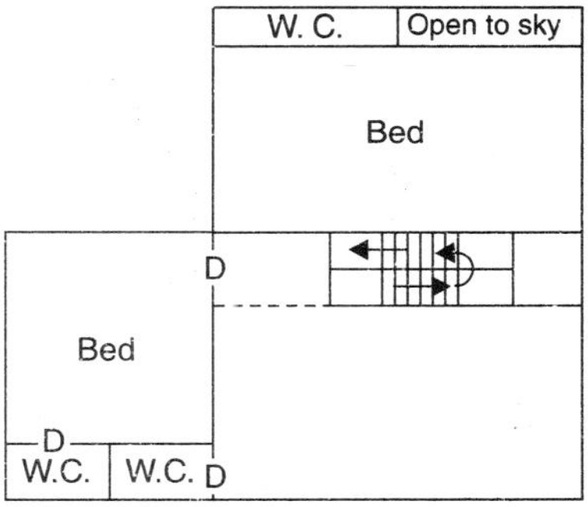

(b) First floor plan

Fig. 9.6 A two storied house

9.4 LAYOUT OF INDUSTRIAL BUILDINGS

The essential components of industrial buildings are:

1. Office
2. Factory
3. Godowns and warehouses.
4. Other facilities.

1. **Office Building:** Every industry needs an office building in the factory area to deal with its own workers and with outside agencies supplying raw materials and purchasing finished project. This building may be independ or attached to factory. It should be close to main entrance to the factory. It should have elegant look to impress outside agencies connected with the business of the industry. It needs rooms for managers, rooms for board meeting and waiting lounge. Office rooms are required for maintaining accounts and establishment records. It is preferable to have two storied building for office, so that it hides the look of factory behind it.

2. **Factory Building:** Requirements of a factory depends upon the manufacturing process involved in the factory. Usually factory buildings are single storey large size buildings to accommodate machineries. Since size of work space is large, the width of factory building is large. To get good lighting and ventilations the walls of factory buildings are 4 to 4.5 m high and roofs are on north light trusses. Flooring is heavy and finishing is not smooth. The area required for machineries is found by drawing sketches, making provision for the movement of men and materials. After completing layouts position of columns and requirement of height of roof is determined. Sufficiently wide doors, windows and ventilators are provided. Areas of such openings should be 30 to 40% of floor area.

3. **Godowns and Warehouses:** Any factory needs at least two stores— one for raw materials and another for finished goods. Raw material store should be close to starting activity of industrial production. It should be accessible for lorries. The store should be connected to factory by issue counters. The size is decided by the materials to be stored and minimum number of days stock required. Finished goods warehouse is close to office and arrangements are made to store the finished goods in attractive manner.

4. Other Facilities: The following facilities are to be provided in the factories:

(*i*) Good entrance
(*ii*) Parking places for vehicles
(*iii*) Cloak room
(*iv*) Drinking water
(*v*) Canteen
(*vi*) Sanitary block
(*vii*) Medical aid
(*viii*) Loading and unloading platforms.

(*i*) *Good entrance:* Since industries work in shift sufficiently wide entrance should be provided for workers to enter and leave. There should be gaurd room for security check. A small showroom/showcase provided at entrance will advertising the items manufactured in the industry.

(*ii*) *Parking place for vehicles:* Infront of entrance cycles stands and two wheelers stands should be provided so that without additional gaurds vehicular movement is observed by gaurds at entrance. Closeness of these places help workers to get down near the entrance to the work place.

(*iii*) *Cloak rooms:* Separate cloak rooms should be provided for ladies and men workers. They should be located inside factory area close to entrance. Cloak rooms should have required numbers of lockers for workers to deposit their valuables and changed clothes.

(*iv*) *Drinking water:* Safe drinking water facility should be provided near cloak rooms and inside the factory. Push button type drinking water taps are preferable to avoid wastage of water.

(*v*) *Canteen:* The workers and sometimes even the officers like to eat the lunch brought from their houses. A dining hall is to be provided in the industry to accommodate workers and officers for this purpose. A separate partitioned area may be provided for officers. Dining tables of different sizes to accommodate 4 to 6 persons should be provided. Kitchen for workers to warm their food and a kitchen for canteen contractor are required. Canteen contractor serves tea, coffee and eatables.

(*vi*) *Sanitary block:* Separate sanitary blocks are required for ladies and gents. Sanitary blocks contain a number of water closets, urinals, wash basins. The size of the blocks depend upon the number of workers. These blocks are provided near cloak rooms and near work places.

(*vii*) *Medical aid:* If any worker is injured while working in the factory, he needs first aid immediately. For this purpose suitable provision should be made in the factory. In bigger industries qualified doctors are appointed for this purpose. There should be arrangement for a vehicle to carry injured to hospitals.

(*viii*) *Loading and unloading platform:* These are provided near raw material store room and finished goods store. Lorries should be able to move conveniently to these platforms and load or unload the materials.

Figure 9.7 shows a typical layout plan of an industry. Figure 9.8 shows the details of office cum factory building. Figure 9.9 shows layout of a typical canteen.

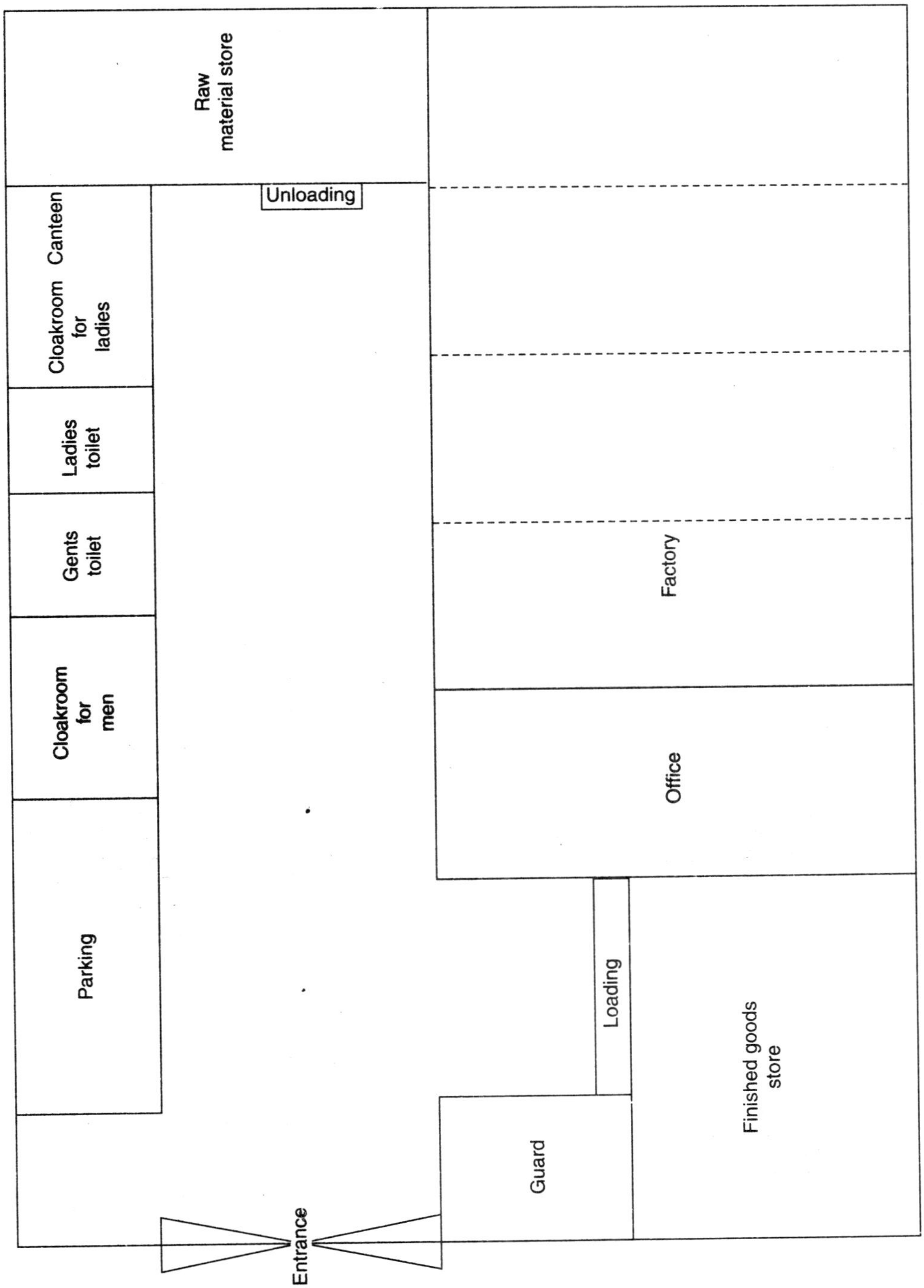

Fig. 9.7 Layout of a factory

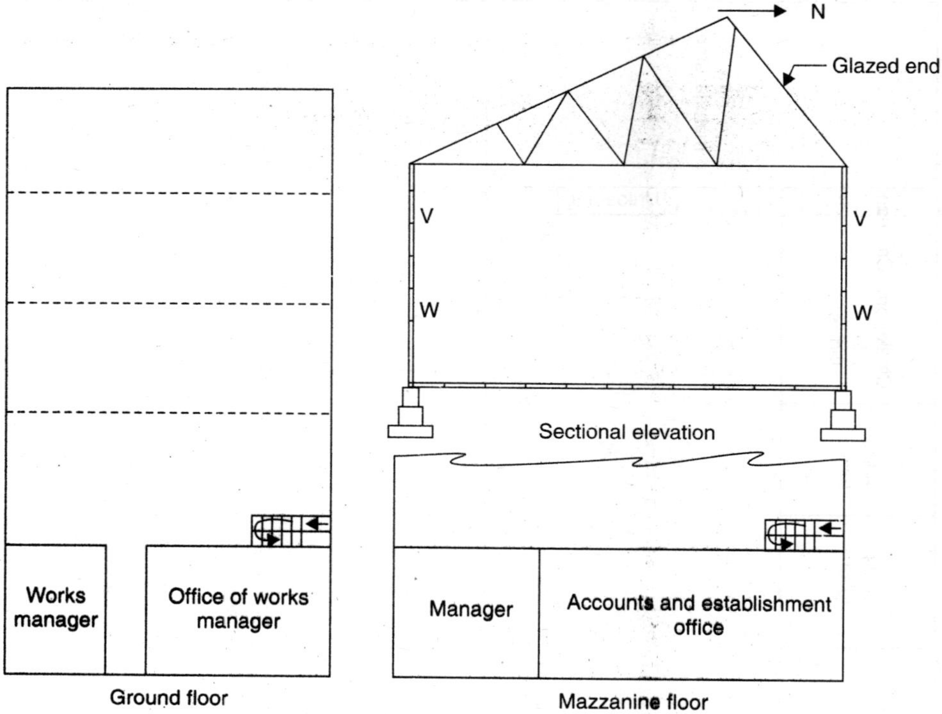

Fig. 9.8 Details of office and factory buildings

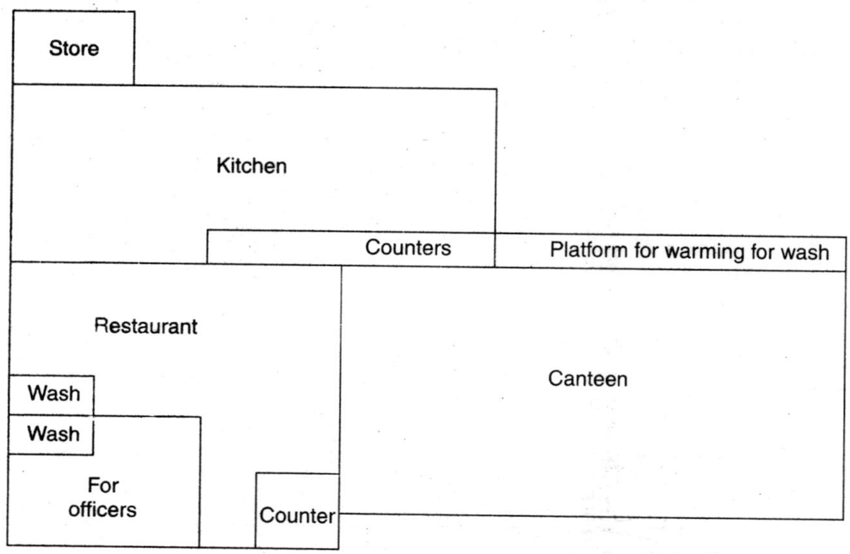

Fig. 9.9 Layout plan of a canteen

9.5 CLASSIFICATION OF BUILDINGS

Buildings may be classified based upon

1. Occupancy
2. Structure

9.5.1 Classification Based on Occupancy

On this basis buildings may be broadly classified as

(*i*) Residential
(*ii*) Educational
(*iii*) Shopping centres
(*iv*) Hotels
(*v*) Banks and offices
(*vi*) Recreation centres
(*vii*) Hospitals and dispensaries
(*viii*) Industrial
(*ix*) Godowns.

There can be further classifications like independent house (bungalows), semidetached, row of houses (chawls) and flats in residential buildings and primary school, high school, college (first grade, engineering, medical, agricultural ...) and universities in case of educational institutions.

9.5.2 Classification Based on Structure

Based on the structure buildings may be classified as

(*i*) Load bearing
(*ii*) Framed

In load bearing constructions, load from roof and floors is transferred to foundations by walls. It is not economical to build multistorey buildings by load bearing construction.

For multistorey building construction, usually framed structures are preferred. In such cases, roof and floor loads are transferred through columns and column footings. Walls serves as only partitions in the building. The walls can be removed and alterations in the building can be carried out without endangering the buildings.

Buildings may be classified on the basis of materials used such as R.C.C. buildings, steel structures, composite structures, etc.

9.6 DESIGN LOADS

To get safe structures at the same time without ignoring economy, it is necessary to estimate various loads likely to act on the structure suitably. Indian standard code IS: 875–1987 specifies various design loads for buildings and structures. They have grouped various loads as under:

1. Dead loads
2. Imposed loads
3. Wind loads
4. Snow loads
5. Earthquake loads and
6. Special loads.

Details of earthquake load is covered in IS: 1893–1984, which should be considered along with other types of loads. The code IS-875 also gives various load combinations to be considered in the design.

9.6.1 Dead Loads

The dead load in a building comprises the weights of roof, floors beams, columns, walls, partition walls etc. which form permanent part of the building. Dead load is found by calculating volume of each part and then multiplying with unit weight of the material used. Unit weights of various materials are presented in part – 1 of IS: 875. Unit weights of some of the common materials are listed in Table 9.1.

Table 9.1 Unit weight of common building materials

Sr. No.	*Material*	*Weight*
1.	Brick Masonry	18.8 kN/m^3
2.	Stone Masonry	20.4 to 26.5 kN/m^3
3.	Plain Concrete	24.0 kN/m^3
4.	Reinforced Cement Concrete	25.0 kN/m^3
5.	Timber	5 to 8 kN/m^3
6.	Wooden Floor with Hard Wood 16 mm thick	0.160 kN/m^2
7.	Terrazo Flooring	0.240 kN/m^2
8.	Country Tiled Roof, including Battens	0.700 kN/m^2
9.	Mangalore Tiled Roof with Battens	0.650 kN/m^2
10.	Mangalore Tiles with Flat Tiles	0.785 kN/m^2
11.	A.C. sheet Roof, 6 mm thick	0.170 kN/m^2
12.	A.C. Sheet Roof, with 5 mm plain	0.110 kN/m^2

9.6.2 Imposed Loads (IL)

The loads which keep on changing from time to time are called as imposed loads. Common examples of such loads in a building are the weight of the persons, weights of movable partitions, dust loads and weight of furnitures. These loads were formerly known as live loads. These loads are to be suitably assumed by the designer. It is one of the major load in the design. The minimum values to be assumed are given in IS 875 (Part 2) – 1987. It depends upon the intended use of the building. These values are presented for square metre of floor area. The code gives the values of loads for the following occupancy classification:

(*i*) Residential buildings—dwelling houses, hotels, hostels, boiler rooms and plant rooms, garages.

(*ii*) Educational buildings

(*iii*) Institutional buildings

(*iv*) Assembly buildings

(*v*) Business and office buildings

(*vi*) Mercantine buildings

(*vii*) Industrial buildings, and

(*viii*) Storage rooms.

The code gives uniformly distributed load as well as concentrated loads. The floors are to be investigated for both uniformly distributed and worst position of concentrated loads. The one which gives worst effect is to be considered for the design but both should not be considered to act simultaneously.

In a particular building, imposed load may change from room to room. For example in a hotel or a hostel building the loads specified are,

	udl	Concentrated load
(*a*) Living rooms and bed rooms	2 kN/m^2	1.8 kN
(*b*) Kitchen	3 kN/m^2	4.5 kN
(*c*) Dining rooms	4 kN/m^2	2.7 kN
(*d*) Office rooms	2.5 kN/m^2	2.7 kN
(*e*) Store rooms	5 kN/m^2	4.5 kN
(*f*) Rooms for indoor games	3 kN/m^2	1.8 kN
(*g*) Bath rooms and toilets	2 kN/m^2	–
(*h*) Corridors, passages, staircases etc.	3 kN/m^2	4.5 kN
(*i*) Balconies	4 kN/m^2	1.5 kN concentrated at outer edge.

Some of the important values are presented in Table 9.2, which are the minimum values and wherever necessary more than these values are to be assumed.

Table 9.2 Minimum imposed load to be considered

Sl. No.	*Occupancy*	*UDL Load*	*Concentrated*
1.	Bathrooms and toilets in all types of building	2 kN/m^2	1.8 kN
2.	Living and bed rooms	2 kN/m^2	1.8 kN
3.	Office rooms in		
	(*i*) Hostels, hotels, hospitals and business building with separate store	2.5 kN/m^2	2.7 kN
	(*ii*) In assembly buildings	3 kN/m^2	4.5 kN
4.	Kitchens in (*i*) Dwelling houses	2 kN/m^2	1.8 kN
	(*ii*) Hostels, hotels and hospitals	3 kN/m^2	4.5 kN
5.	Banking halls, classrooms, x-ray rooms, operation rooms	3 kN/m^2	4.5 kN
6.	Dining rooms in (*i*) educational buildings, institutional and mercantine buildings	3 kN/m^2	2.7 kN
	(*ii*) hostels and hotels	4 kN/m^2	2.7 kN
7.	Corridors, passages, staircases in		
	(*i*) Dwelling houses, hostels and hotels	3 kN/m^2	4.5 kN

	(*ii*) Educational institutional and assembly buildings	4 kN/m^2	4.5 kN
	(*iii*) Marcantine buildings	5 kN/m^2	4.5 kN
8.	Reading rooms in libraries		
	(*i*) With separate storage	3 kN/m^2	4.5 kN
	(*ii*) Without separate storage	4 kN/m^2	4.5 kN
9.	Assembly areas in assembly buildings		
	(*i*) With fixed seats	5 kN/m^2	..
	(*ii*) Without fixed seats	5 kN/m^2	3.6 kN
10.	Store rooms in educational buildings	5 kN/m^2	4.5 kN
11.	Store room in libraries	6 kN/m^2 for a height of 2.24 + 2 kN/m^2 for every 1m additional height	4.5 kN
12.	Boiler rooms and plant rooms in		
	(*i*) Hostels, hotels, hospitals, mercantine and industrial buildings	5 kN/m^2	4.5 kN
	(*ii*) Assembly and storage buildings	7.5 kN/m^2	4.5 kN

Imposed loads to be considered on various roofs are presented in Table 9.3.

Table 9.3 Imposed loads on various types of roofs
(Table 2 of National Building Code - 1983)

Sl. No.	*Type of roof*	*Imposed load measured on plan area*	*Minimum imposed load measured on plan*
(*i*)	Flat, sloping or curved roof with slopes up to and including 10 degrees		
	(*a*) Access provided	1.5 kN/m^2	3.75 kN uniformly distributed over any span of one metre width of the roof slab and 9 kN uniformly distributed over the span of any beam or truss or wall.
	(*b*) Access not provided except for maintenance	0.75 kN/m^2	1.9 kN uniformly distributed over any span of one metre width of the roof slab and 4.5 kN uniformly distributed over the span of any beam of truss or wall.

(*ii*)	Sloping roof with slope greater than 10 degrees	For roof membrane sheets or purlins – 0.75 kN/m^2 less 0.02 kN/m^2 for every 10 degrees	Subject to a minimum of 0.4 kN/m^2
(*iii*)	Curved roof with slope of line obtained by joining springing point to the crown with the horizontal, greater than 10 degrees	$(0.75 - 0.52\ \alpha^2)$ kN/m^2 where $\alpha = h/l$ h = height of the highest point of the structure measured from its springing: and l = chord width of the roof if singly curved and shorter of the two sides if doubly curved. Alternatively, where structural analysis can be carried out for curved roofs of all slopes in a simple manner applying the laws of statistics, the curved roofs shall be divided into minimum 6 equal segments and for each segment imposed load shall be calculated appropriate of each segment as given in (*i*) and (*ii*)	Subject to a minimum of 0.4 kN/m^2

Note: 1. The loads given above do not include loads due to snow, rain, dust collection, etc. The roof shall be designed for imposed loads above or snow/rain load, whichever is greater.

Note: 2. For special types of roofs with highly permeable and absorbent material, the contingency of roof material increasing in weight due to absorption of moisture shall be provided for.

However in a multi-storeyed buildings chances of full imposed loads acting simultaneoulsy on all floors is very rare. Hence the code makes provision for reduction of loads in designing columns, load bearing walls, their supports and foundations as shown in Table 9.4.

Table 9.4 Reductions in imposed loads on floors in design of supporting structural elements

Number of floors (including the roof) to be carried by member under consideration	*Reduction in total distributed imposed load in per cent*
1	0
2	10
3	20
4	30
5 to 10	40
Over 10	50

9.6.3 Wind Loads

The force exerted by the horizontal component of wind is to be considered in the design of buildings. It depends upon the velocity of wind and shape and size of the building. Complete details of calculating wind load on structures are given in IS-875 (Part 3) - 1987. Brief idea of these provisions are given below:

(*i*) Using colour code, basic wind pressure 'V_b' is shown in a map of India. Designer can pickup the value of V_b depending upon the locality of the building.

(*ii*) To get the design wind velocity V_z the following expression shall be used:

$$V_z = k_1\, k_2\, k_3\, V_b$$

Where k_1 = Risk coefficient

k_2 = Coefficient based on terrain, height and structure size.

k_3 = Topography factor

(*iii*) The design wind pressure is given by

$$p_z = 0.6\, V_z^2$$

where p_z is in N/m^2 at height Z and V_z is in m/sec. Up to a height of 30 m, the wind pressure is considered to act uniformly. Above 30 m height, the wind pressure increases.

9.6.4 Snow Loads

IS 875 (Part 4) – 1987 deals with snow loads on roofs of the building. For the building to be located in the regions wherever snow is likely to fall, this load is to be considered. The snow load acts vertically and may be expressed in kN/m^2 or N/m^2. The minimum snow load on a roof area or any other area above ground which is subjected to snow accumulation is obtained by the expression

$$S = \mu\, S_0$$

Where S = Design snow load on plan area of roof.

μ = Shape coefficient, and

S_0 = Ground snow load.

Ground snow load at any place depends on the critical combination of the maximum depth of undisturbed aggregate cumulative snow fall and its average density. These values for different regions may be obtained from Snow and Avalanches Study Establishment, Manali (HP) or Indian Meteorological Department, Pune. The shape coefficient depends on the shape of roofs and for some of the common shapes the code gives these coefficients. When the slope of roof is more than 60° this load is not considered.

It may be noted that roofs should be designed for the actual load due to snow or for the imposed load, whichever is more severe.

9.6.5 Earthquake Forces

Earthquake shocks cause movement of foundation of structures. Due to inertia forces develop on super structures. The total vibration caused by earthquake may be resolved into three mutually perpendicular directions, usually taken as vertical and two horizontal directions. The movement in vertical direction do not cause forces in superstructure to any significant extent. But movement in horizontal directions cause considerable forces.

The intensity of vibration of ground expected at any location depends upon the magnitude of earthquake, the depth of focus, the distance from the epicenter and the strata on which the structure stands.

The response of the structure to the ground vibration is a function of the nature of foundation, soil, size and mode of construction and the duration and intensity of ground motion. IS: 1893–1984 gives the details of such calculations for structures standing on soils which will not considerably settle or slide appreciably due to earthquake. The seismic accelerations for the design may be arrived at from **seismic coefficients,** which is defined as the ratio of acceleration due to earthquake and acceleration due to gravity. For the purpose of determining the seismic forces, India is divided into five zones. Depending on the problem, one of the following two methods may be used for computing the seismic forces:

(*a*) Seismic coefficient method (*b*) Response spectrum method

The details of these methods are presented in IS 1983 code and also in National Building Code of India. After the Gujarat earthquake (2000) Government of India has realised the importance of structural designs based on considering seismic forces and has initiated training of the teachers of technical institution on a large scale (NPEEE).

There are large number of cases of less importance and relatively small structures for which no analysis be made for earthquake forces provided certain simple precautions are taken in the constructions. For example

(*i*) Providing bracings in the vertical panels of steel and R.C.C. frames.

(*ii*) Avoiding mud and rubble masonry and going for light materials and well braced timber framed structures.

9.6.6 Other Forces and Effects

As per the clause 19.6 of IS 456-2000, in addition to above load discussed, account shall be taken of the following forces and effects if they are liable to affect materially the safety and serviceability of the structure:

(*a*) Foundation movement (See IS 1904)

(*b*) Elastic axial shortening

(*c*) Soil and fluid pressure (See IS 875, Part 5)

(*d*) Vibration

(*e*) Fatigue

(*f*) Impact (See IS 875, Part 5)

(*g*) Erection loads (See IS 875, Part 2) and

(*h*) Stress concentration effect due to point load.

9.6.7 Load Combinations

A judicious combination of the loads is necessary to ensure the required safety and economy in the design keeping in view the probability of

(*a*) their acting together

(*b*) their disposition in relation to other loads and severity of stresses or deformations caused by the combination of various loads.

The recommended load combinations by codes are

1.	DL	7.	DL + IL + EL
2.	DL + IL	8.	DL + IL + TL
3.	DL + WL	9.	DL + WL + TL
4.	DL + EI	10.	DL + EL + TL
5.	DL + TL	11.	DL + IL + WL + TL
6.	DL + IL + WL	12.	DL + IL + EL + TL

where DL = dead load IL = imposed load

WL = wind load EL = earthquake load

and TL = temperature load.

Note: When snow load is present on roofs, replace imposed load by snow load for the purpose of above load combinations.

9.7 COMMON BUILDING COMPONENTS, THEIR FUNCTIONS AND NOMINAL DIMENSIONS

The following are the common building components:

1. Foundation
2. Plinth
3. Walls and columns
4. Sills, lintel and chejjas
5. Doors and windows
6. Floors
7. Roof
8. Steps, stairs and lifts
9. Finishing work
10. Building services.

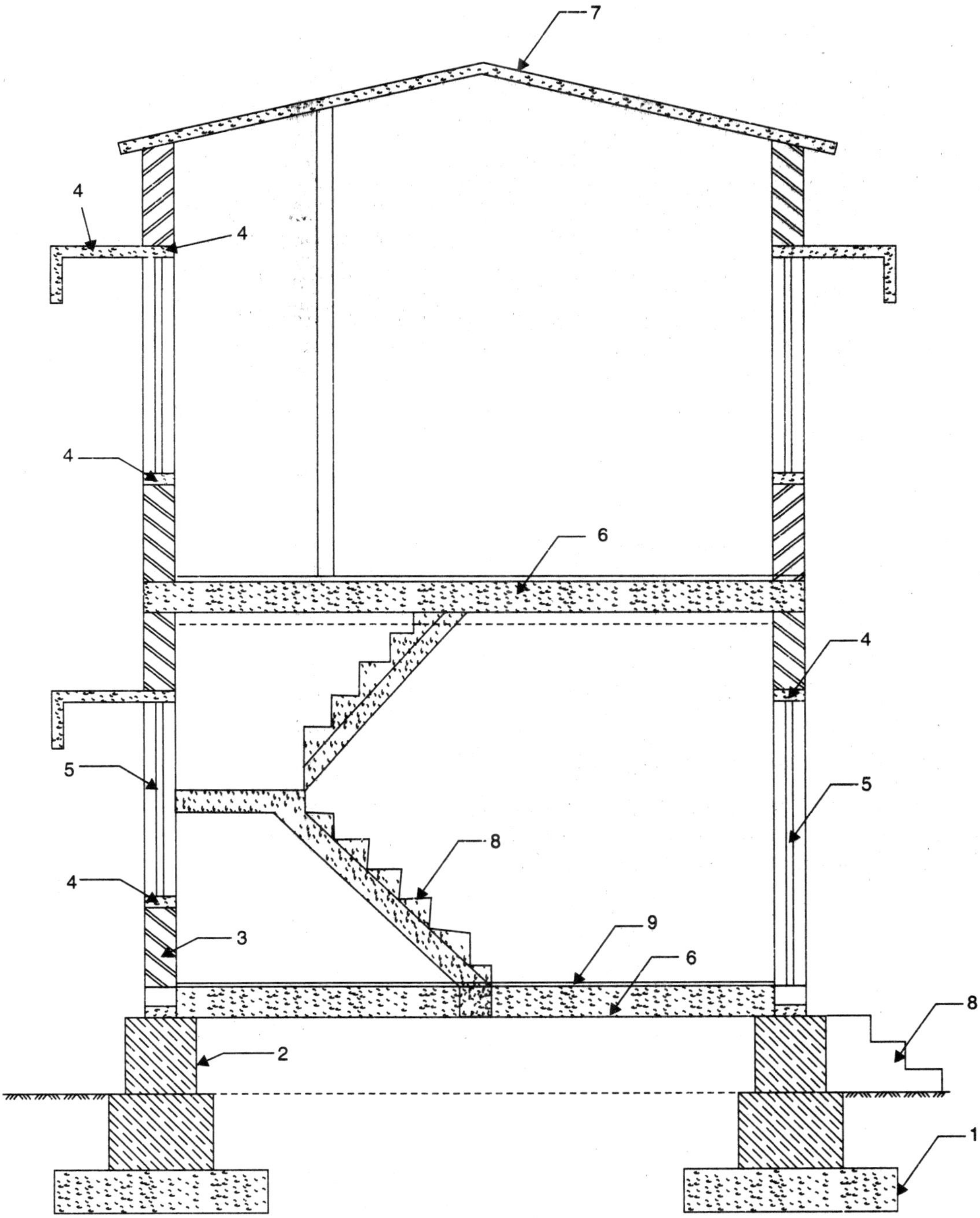

1. Foundation	2. Plinth	3. Walls and columns
4. Sills, lintels and chejjas	5. Doors and windows	6. Floors
7. Roof	8. Steps and stairs	9. Finishing work

Fig. 9.10 Components of a building

Functions and Nominal Dimensions of Component of Buildings:

9.7.1 Foundation

It is the most important part of the building. Building activity starts with digging the trenches for foundation and then building it. It is the lower most part of the building. Its *main functions are:*

1. Distribute the load from the structure on soil evenly and safely.
2. To anchor the building to the ground so that under lateral loads building will not move.
3. It prevents the building from overturning due to lateral forces.
4. It gives even surface for the construction of super structure.

Minimum Dimensions:

(*a*) *Depth of foundation:* For all types of foundation minimum depth of foundation is calculated using Rankine's formula:

$$H = \frac{p}{w}\left(\frac{1-\sin\phi}{1+\sin\phi}\right)^2$$

where p = safe bearing capacity of soil

w = unit weight of soil

ϕ = angle of repose of soil.

However in any case it is not less than 0.9 m.

Finding safe bearing capacity of a soil is an experts job and it is found after conducting the tests in field/laboratories. However general values for common soils are listed in Table 2 of National Building Code.

(*b*) *Width of foundation:* Width of wall foundations or size of column foundation is determined by first calculating the load on foundation and then dividing that with safe bearing capacity: Thus

$$\text{Width of wall foundation} = \frac{\text{Load per unit length of wall}}{\text{Safe bearing capacity of soil}}.$$

$$\text{Area of column footing} = \frac{\text{Load carried by column}}{\text{Safe bearing capacity of soil}}$$

The wall foundation width is reached by gradually increasing the width of courses in foundation. The projection from course to course in foundation is 75 mm on either side. Finally plain concrete bed can have a projection of 150 to 200 mm on either side.

Foundations for footing are usually of R.C.C. which are designed to resist upward thrust by soil.

9.7.2 Plinth

The portion of the wall between the ground level and the ground floor level is called plinth. It is usually of stone masonry. If the foundation is on piles, a plinth beam is cast to support wall above floor level. At the top of plinth a damp proof course is provided. It is usually 75 mm thick plain concrete course.

The function of plinth is to keep the ground floor above ground level, free of dampness. Its height is not less than 450 mm. It is required that plinth level should be at least 150 mm above the road level, so that connections to underground drainage systems can be made.

9.7.3 Walls and Columns

The function of walls and columns is to transfer the load of structure vertically downwards to take it to ground. Apart from this, wall performs the following functions also:

1. It encloses building area into different compartments and provides privacy.
2. It provides safety to inmates from burglary and insects.
3. It keeps building warm in winter and cool in summer.

The multistorey buildings are provided with R.C.C. columns or composites of steel and concrete. The sizes of columns are decided by the loads acting and load carrying capacity of the materials of the columns.

Minimum Thickness: Stone or brick masonry walls are built course by course in 1 : 6 cement mortars. The height of a course is 150 mm in stone masonry and 100 mm in case of brick masonry. The thickness of wall should be sufficient to transfer the load safely, without exceeding the permissible stress. The critical portions in masonry from consideration of stresses are near the openings of doors and windows and the positions where concrete beams rest.

Minimum thicknesses used are 375 mm, 200 mm and 200 mm in case of stone, brick and hollow block constructions respectively. It is also recommend that slenderness ratio of wall defined as ratio of effective length or effective height to thickness should not be more than 27. National code of India, defines effective height and effective length as given in Table 9.5 and 9.6 [For full details refer NBC1983].

Table 9.5 Effective height of walls in terms of actual height H

Sr. No.	*End condition*	*Effective height*
1.	Lateral as well as rotation restraint	0.75 H
2.	Lateral as well rotational restraint at one end and only lateral restraint at other	0.85 H
3.	Lateral restraint but no rotational restraint at both ends	1.0 H
4.	Lateral and rotational restraint at one end and no restraint at other ends (compound walls, parapet walls etc.)	1.5 H

Table 9.6 Effective length of walls of length L

Sr. No.	*Condition of support*	*Effective length*
1.	Continuous and supported by cross walls	0.8 L
2.	Continuous at one end and supported by cross wall at the other end	0.9 L
3.	Wall supported by cross walls at each end	1.0 L
4.	Free at one end and continuous at other end	1.5 L
5.	Free at one end and supported by cross wall at other end	2.0 L

However it may be noted as a guide line that up to 3 storey building 200 mm thick brickwall is required. If it is 4 storey building a minimum thickness of 300 mm should be used in the ground floor and the rest may be with 200 mm thickness.

9.7.4 Sills, Lintels and Chejjas

A window frame should not be directly placed over masonry. It is placed over 50 to 75 mm thick plain concrete course provided over the masonry. This concrete course is called sill. Lintels are the R.C.C. or stone beams provided over the door and window openings to transfer the load transversely so as to see that the window or door frame is not unduly stressed. The width of such beams is equal to the width of wall while thickness to be provided depends upon the opening size. Suitable calculations are required to determine the thickness so as to keep the stresses below permissible values. A minimum of 100 mm thickness is provided for all R.C.C. lintels. Chejja is the protection given outside the wall to protect doors and window from rain. They are usually made with R.C.C. In low cost houses some time stone slabs are provided as chejjas. The minimum thickness provided on the face of the wall is 75 mm in case of R.C.C. chejjas. The projection varies from 600 mm to 800 mm. Sometimes drops are also provided to chejjas to improve aesthetic look and also to get additional protection from sun and rain.

9.7.5 Doors and Windows

The function of a door is to give access to different rooms in the building and to deny the access whenever necessary. Number of doors should be minimum possible. The size of the door should be of such dimension as will facilitate the movement of the largest object likely to use the door.

In case of residential buildings the size of the door should not be less than 0.9 m × 2.0 m. Larger doors may be provided at entrance to building to enhance the aesthetic view. Minimum size doors are used for bath rooms and water closets. The size recommended is 0.75 m × 1.90 m. As a thumb rule height of door should be 1.0 m more than the width.

Windows are provided to give light and ventilation. They are located at a height of 0.75 m to 0.90 m from the floor level. In hot and humid regions, the window area should be 15 to 20 per cent of the floor area. It is preferable to have at least two openings in two different walls. Another thumb rule used to determine size of the window opening is for every 30 m^3 inside volume there should be 1 m^2 window opening.

9.7.6 Floors

Floors are important component of a building. They give working/useful area for the occupant. A ground floor is prepared by filling brick bats, waste stones, moorum and well competed with not less than 100 mm sand layer on its top. A lean concrete of 1 : 4 : 8, 100 mm thick is laid. A damp proof course may be provided to prevent capillary rise of water from ground. Then floor finishing is done as per the requirement of the owner. Cheapest floor finish for a moderate house is with 20 to 25 mm rich mortar course, finished with red oxide. The costliest floor finish is mossaic or marble flooring.

Other floors are usually R.C.C. floors. The finishing of the floor should be as per the requirements of the owner. Floor finish adds to the aesthetic of the building.

9.7.7 Roof

Roof is the top most portion of the building which provide top cover to the building. It should be leak proof.

Sloping roofs like tiled roof, and A.C. sheet roof give leak proof cover easily. But they do not give provision for constructing additional floor, if required in future. Tiled roof give protection from sun also, but A.C. sheet roofs get heated up and result in rise in temperature of the building. Many may not like sloping roof for their aesthetic view.

Flat roofs give provision for additional floors. The terrace adds to the comfort of occupants. Water tanks can be easily placed over flat roofs. Modern flat roofs are mainly of R.C.C. Rain water should be drained out quickly by giving suitable slopes. Stagnation of water on any part of flat roof causes leakage and creates unhygenic condition and reduces the life of the structure. In R.C.C. flat roof a minimum of 120 mm thick slab is to be used. The slab may need beams, which is decided by R.C.C. designer.

9.7.8 Steps, Stairs and Lifts

Steps give convenient access from ground level to ground floor level. They are required at doors in outer walls. 250 to 300 mm width and a rise of 150 mm is ideal step size. In no case rise of two consecutive steps should be different. Number of rises required depends upon the difference in the levels of ground and the ground floor.

Stairs give access from floor to floor. The following are the recommended layout of a stair case:

1. **Width:** 0.9 m in residential buildings, 1.5 to 2.0 m in public buildings.
2. **Number of steps in a flight:** Maximum number in a flight should be restricted to 12 to 14 while minimum is 3.
3. **Rise:** Rise provided in a residential building is 150 mm to 175 mm whereas in public buildings it is 120 mm to 150 mm. However in commercial buildings, rises are kept higher from the consideration of economic floor area.
4. **Tread:** For residential buildings tread or going is 250 mm, while in public buildings preferable tread is 270 mm to 300 mm.

The dimensions of rise and tread should not be changed in any stair. The following empirical formula is used to decide rise and tread:

$$2R + T > 550 \text{ mm} \quad \text{but} \quad < 700 \text{ to } 600 \text{ mm.}$$

Head Room: Head room available in a stair case should not be less than 2.1 m.

Hand Rails: Hand rails should be provided at a convenient height of a normal person, which is 850 mm to 900 mm.

Lifts: In all public buildings lifts are to be provided for the convenience of old and disabled persons. In hostels G + 3 floors can be built without lift, but in residential flats out G + 2 floors are to be built without lifts. For all multistorey buildings lifts are required.

Lift is to be located near the entrance. Size of the lift is decided by the number of users in peak hours. Lifts are available in the market with capacity for 4 to 20 person, with average weight of a person considered as 68 kg. IS 14665 parts I to IV give specification for lifts.

9.7.9 Finishing

It consists in giving finishing touches to bottom portions of slab, wall surface and top of floor. Plastering is the usual finishing touch to lower portion of slab and walls. Painting, distempering or decorative colour washing is applied. The function of finishing work is

(*i*) They give protective layer

(*ii*) They improve aesthetic appearance.

(*iii*) They rectify defective workmanship.

Finishing work for plinth consists in pointing the joints.

Finishing work for floor is polishing it.

9.7.10 Building Services

Water supply, drainage, sanitation, construction of cupboards, showcases and electric supply are considered as building services.

For storing water from municipal supply or from tanker supply a sump is built in the house property near street. Sump capacity can be as high as 9 m^3 to 12 m^3, so that supply from a tanker is stored. From the sump water is pumped to over head tanks placed on or above roof level, so as to get water inside the house all the 24 hours. Necessary arrangements for this purpose is to be made. Plumbing work is done so as to get water in kitchen, bath rooms, water closets, sinks and garden taps.

Service pipes of 20 mm are used. For all other connections 12 mm pipes are used. Many people prefer concealed piping for aesthetic point of view. In such cases care should be taken to properly seal all joints. To avoid corrosion problem P.V.C. pipes are preferred for concealed piping.

For draining rain water from roofs, down take pipes of at least 100 mm diameters should be used. Proper slopes should be given to roof towards down take pipes. These pipes should be fixed at 10 to 15 mm below the roof surface so that water is directed to pipe easily.

Sanitary fittings of required colours may be provided in the water closets. They are to be connected to stoneware pipes with suitable traps in between. Stoneware pipes are then connected to UGD (under ground drainage) of municipal lines or to septic tank.

Many carpentary works are required for building service. They are in the form of showcases, cupboards, racks, etc.

Electricity supply is essential part of building services. The building should be provided with sufficient points for safe supply for lights, fans and other electric gadgets.

Usually separate estimations are made for building services. It roughly comes out to be 7.5 per cent of building cost for water supply and sanitation and another 7.5 per cent for electric supply.

9.8 ELEMENTS OF BUILDING DRAWING

Drawing is the language of engineers. The details of building are expressed with more clarity by drawings rather than with words. Apart from site plan, building drawing consists of at least the following three views:

1. Plan
2. Elevation
3. Section.

All drawings are made to scale and scale is clearly indicated in drawing. Many municipalities insist for standard scales like 1 : 50 or 1 : 100. Site plans are drawn to 1 : 200 scale and location plans to 1 : 1000.

9.8.1 Plan

Plan is the view from top. Normally sectional plans are drawn to show details. The section imagined is slightly above the sill level so that details of window positions are also shown. Any important details above this level, it must be shown, they are shown by dotted lines; *e.g.*, position of a beam or position of steps. Plan gives the following details:

(*a*) Inner dimensions of all rooms and halls. If a room dimension is written as 3600 × 4000 it means right to left dimension is 3600 mm and at right angles it is 4000 mm.

(*b*) Thicknesses of walls, partition walls, etc.

(*c*) Dimensions of columns.

(*d*) Position of doors and windows. It also indicates whether door window frame should be flush with outer surface of wall or in the middle of wall or slightly inside the wall.

(*e*) It indicates plinth projection, chejja projection etc.

If arrangements of rooms is different in different floors sectional plans should be drawn for each floor separately.

9.8.2 Elevation

Elevation is the front view of the building. Ground line and the view of building above that are drawn to scale. Projections are made from the plan to get the position of salient points in elevation. No dimensions are written in the elevation. It gives aesthetic view of the building from road side. This may be of more importance for the owner of the building than for engineers or construction mistry.

9.8.3 Section

Sectional elevations are drawn to show heights of various objects in building. First a sectional line is drawn in plan view across which is to be drawn. This line need not be a single straight line. It can be shifted parallel to itself in some portion to show important details. In sectional views the following details are shown.

(*a*) Heights of various courses of foundation with details of materials, indicated. Type of material is indicated by drawing conventions. In case of concrete, mix proportion is also written.

(*b*) Thickness of damp proof course, thicknesses of flooring and type of floor finish indicated. The other details shown are

(*c*) Thickness of walls.

(*d*) Position of door and window frames with respect to wall.

(*e*) Rise and tread of stairs, steps etc.

(*f*) Thicknesses of slabs, chejjas etc.

(*g*) Depth of beams and lintels.

(*h*) Height of room.

(*i*) Height of sill level and thickness of sill concrete.

(*j*) Height of parapets, hand rails.

If necessary more number of sectional views are to be drawn. All drawings should be indexed giving details of size and type of doors, windows and ventilators.

Schedule giving details of site number, name of area and city, owners name and scale should be shown in the right hand side lower corner.

Structural drawings, giving the details of R.C.C. structural members like slab, beam, columns and footings are separately drawn. Apart from giving size of the members, they give reinforcement details.

9.9 INTRODUCTION TO BUILDING BYE-LAWS

To prevent hapazard development and ensure orderly development Indian Standards have been prepared and published as IS 1256–1967 as code of building bye-laws. Municipalities and other competant authorities have power to make changes in these bye-laws to suit their local requirements. These bye-laws are necessary to achieve the following objectives:

1. To allow orderly growth and prevent hapazard developments.
2. To afford safety against fire, noise, health hazard and structural failures.
3. To give guidelines to owners, engineers and architects to ensure minimum standards. Bye-laws should be respected by every one. The plan sanctioned by competant authority should not be changed violating bye-laws. In such cases the competant authority is having every right to demolish violated portions. Before going through byelaws it is necessary to understand the following terminologies:

Building Height: It is the vertical distance between the centre line of street infront of the building and the highest point of the building. Features provided only for architectural purposes should not be taken for calculating height of the building. In case of flat roofs the measurement of height is up to the top of parapet. In case of pitched roofs building height is considered up to the point where external wall touches sloping roof. In case of the gable facing the street it is considered up to the mid point of gable end [Fig. 9.11].

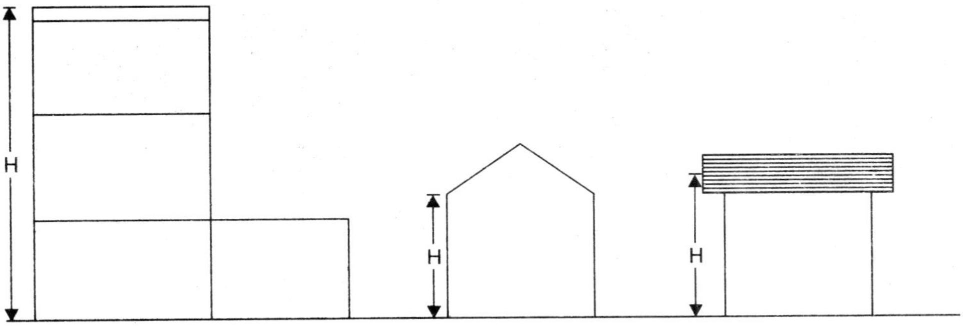

(a) Flat roof (b) Sloped roof sloping towards street (c) Cable end towards street

Fig. 9.11

Building Line: It is the horizontal distance of the permissible building point from the centre line of street. No portion of building should be at less than this distance from the centre of the street infront of the building. This is fixed by competent authority, taking care for future widening of the road, safe distances from electric line etc.

Set backs: It is line parallel to the boundaries of the flat beyond which building cannot be constructed towards site boundaries set backs are prescribed on front, side and rear boundaries of the site.

Covered or Built-up Area: The area covered by the building immediately above the plinth level is called covered or built up area. Covered area do not include the following:

(*i*) Well structure (*ii*) Storage tank

(*iii*) Open verandah, portico (*iv*) Ramp

(*v*) Watchman booth, compound wall (*vi*) Uncovered staircase.

Floor Area Ratio (FAR): It is the ratio obtained by dividing covered area of all the floors with area of plot. Thus,

$$\text{FAR} = \frac{\Sigma \text{ Covered area of floors}}{\text{Area of plot}}$$

Floor Area: Floor area is having different meaning compared to covered area. Floor area of a room is the area of floor between the walls of the room. Floor area of a building is the sum of floor areas of all the rooms, verandahs, passages, kitchen, bath rooms, water closets, staircase, etc. In other word this area is equal to the plinth area minus area occupied by walls and columns.

Sanctioned Plan: Before starting any building construction, the building drawings should be submitted to municipality or any other competant authority of the area. The building plan should be got prepared from a licenced architect or consulting engineer, who will prepare it without violating building byelaws. Once the plan is approved, it becomes sanctioned plan. The owner has to build the house as per sanctioned plan. Minor internal changes do not require permission again but major changes, expansion, etc. need the permission.

Building Bye-laws: The following are the some of important bye-laws:

(*a*) *FAR:* Permissible FAR are fixed by municipal administration or development authorities for different types of the building:

Table 9.7

Sr. No.	*Use*	*Area*	*FAR*	*Remarks*
1.	Residential	Scheme area	1	Maximum allowed on GF = 0.40
2.	Residential	City area	4	Facing street of more than 12 m width.
3.	Residential	City area	3	Facing street of less than 12 m width.
4.	Industrial	Scheme are	1.50	Maximum on GF = 1.0
5.	Commercial	Scheme area	1.33	Maximum on GF = 0.4

(*b*) *Distance From Power Supply Main:* To safeguard from mishaps of electrocution, the distance between any accessible part of the building (balcony, terrace, etc.) and electric supply mains should have the following vertical and horizontal distances (Table 9.8):

Table 9.8

Sr. No.	*Voltage of electric lines*	*Minimum distance*	
		Vertical	*Horizontal*
1.	Low-Medium	2.5 m	1.2 m
2.	High Voltage		
	(*a*) Up to 11 kV	3.7 m	1.2 m
	(*b*) More than 11 kV	3.7 m	2.0 m

(*c*) *Exterior Open Spaces:* From the consideration of lighting and ventilation, NBC recommends minimum open spaces around a building up to 10 m height as given below:

(*i*) *Front Open Space:* It is as shown in Table 9.9 below:

Table 9.9

Sr. No.	*Width of street in fronting the plot*	*Minimum front open space*	*Remarks*
1.	Up to 7.5 m	1.5 m	For buildings up to a height of 7 m
2.	7.5 m to 18.0 m	3.0 m	For buildings up to a height of 10 m
3.	18.0 m to 30.0 m	4.5 m	For buildings up to a height of 10 m
4.	Above 30 m	6.0 m	For buildings up to a height of 10 m

Note: For sites abutting two or more streets the value of open space is based on the average width of street, but in case it is less than 1.8 m.

(ii) *Rear Open Space:* Minimum rear open space should be an average of 3.0 m and in no place it is less than 1.8 m. In case of back to back space, rear open space should be a minimum of 3.0 m throughout. For plots of depth less than 9 m, for building of height less than 7.0 m, minimum is reduced to 1.9 m.

(iii) *Side Open Space:* For detached buildings, minimum side open space recommended is 3.0 m on each side and 3.0 m on only side for semi detached houses. However for smaller plots it may be reduced to 1.5 m by competent authority.

For buildings with above 10 m heights, open space provisions are as shown in Table 9.10.

Table 9.10

Height of building	*Minimum open space requirement*
10 m to 25 m	Minimum for 10 m high building + 1 m for every 3 m additional height.
25 to 30 m	Minimum open space of 30 m.
More than 30 m	30 m + 1 m for every additional 5 m height subject to a maximum of 16 m.

(d) *Height of Building:* Maximum height of a building is restricted according to the width of the street also.

(i) It should not be more than 1.5 times the distance from centre of the road from the front line of building.

(ii) In the vicinity of aerodromes, the minimum height is fixed in consultation with Civil Aviation authorities.

(e) *Plinth Area:* In a market place it should not exceed 75% of the site, provided sufficient street parking facility is given in the plot for loading and unloading.

In industrial area it is restricted to 60% of plot area. For residential area the plinth area bye-laws are as given below in Table 9.11.

Table 9.11

Sr. No.	*Area of plot*	*Maximum permissible plinth area*
1.	< 200 m^2	$66\frac{2}{3}$% of plot area.
2.	201% to 500 m^2	50% of plot area or 133 m^2 whichever is more.
3.	501 to 1000 m^2	40% of plot area or 250 m^2 whichever is more.
4.	> 1000 m^2	$33\frac{1}{3}$% of the plot area or 400 m^2 whichever is more.

Requirements of Parts of Buildings

1. *Plinth Height:* It should be such that there is no difficulty in the drainage. It should be 0.3 m above.

(*i*) centre of the street

(*ii*) foot path of abutting street

(*iii*) the highest part of a service line.

Every interior courtyard should be atleast 150 mm above the nearest street.

2. *Habitable Room:* Habitable rooms in a building should have to satisfy the minimum requirements as shown in Table 9.12.

Table 9.12

	Building with single habitable room	*Building with two habitable rooms*	
		1st room	*2nd room*
Min. floor area	9.5 m^2	9.5 m^2	7.5 m^2
Min. width	2.4 m	2.1 m	2.1 m
Min. height	2.75 m	2.75 m	2.75 m
Min. head room	2.4 m	2.4 m	2.4 m

3. *Kitchen:* National Building Code (NBC) recommends the following:

(*a*) At least one wall of kitchen should abut the exterior open space.

(*b*) The requirements shown in Table 9.13 should be full filled:

Table 9.13

	Min. floor area	*Min. width*	*Min. height*
Kitchen cum store	5.0 m^2	1.8 m	2.75 m
Kitchen with store	4.50 m^2	1.8 m	2.75 m
Kitchen cum dining	7.5 m^2	2.10 m	2.75 m

4. *Bathrooms and Water Closets:* At least one wall of bath, W.C. should abut the exterior open space. At least one bath and W.C. or bath cum W.C. should be provided for every 6 member of users. High level window or ventilator should be provided. The bathroom should be provided with an impervious floor sloping towards drain. Wall should be finished with watertight plaster or with glazed tiles to atleast 1.2 m height. Other requirements are as specified in Table 9.14.

Table 9.14

	Bath	*W.C.*	*Bathroom W.C.*
Minimum floor area	1.80 m^2	1.10 m^2	2.80 m^2
Minimum height	2.0 m	2.0 m	2.0 m
Recommended size	1.5 × 1.2 m	1.10 × 1.0 m	2.3 × 1.2 m

5. *Store Room:* Recommendation for store room in a residential building are:

Minimum area = 3.0 m^2

Nominal size = 1.8 m × 2.2 m

Height = 2.5 m

6. *Stair Case:* For a residential building.

Maximum rise = 190 mm

Minimum tread = 250 mm

Minimum width of stairs = 0.75 m

Maximum number of steps in a flight = 15

Minimum headroom = 2.2 m

Height of hand ra 1 = 0.8 m to 1.0 m.

In public buildings

Min. width = 1.2 m

7. *Compound Wall:* Maximum height above centre line of street = 1.5 m

However a compound wall up to a height of 2.4 m is permitted but top 0.9 m wall should be of open construction like with welded steel grill or R.C. jalis.

In case of corner plots, compound wall should be rounded off. Up to lower 0.75 m may be of solid masonry but top 0.75 m it should be of open type at least for a distance of 10 m from the intersection point.

Submission of Drawing for Issue of Construction Permission

For building construction, permission should be obtained in writing from competent authority. The application for permission should include the following drawings:

1. Site plan, 1 : 1000 scale.
2. Building plan, 1 : 100 scale
3. Service plan, 1 : 200 scale
4. Area statement,
5. Schedule of openings
6. Specifications
7. Stability certificate.

All these documents should be signed by a registered architect/engineer/licensed engineer/licensed supervisor.

Plate 9.1 shows such drawing.

After completing the work, completion certificate should be obtained from the competent authority, before using the building.

QUESTIONS

1. Explain elementary principles of building planning.
2. List various basic requirements of a building and briefly explain them.
3. Explain essential components of a residential building.
4. Write detailed note on layout of industrial buildings.
5. List the classification of buildings based on occupancy and type of structures.
6. Briefly explain the different types of loads acting on a building. Which are the authorities specifying these loads ?
7. What are the functions of a foundation ? On what basis its minimum dimension is determined ?
8. Write short notes on the functions and minimum dimensions of the following building components:

   (*a*) Wall (*b*) Column

   (*c*) Doors and windows (*d*) Steps and stairs.
9. Explain the following terms:

   (*a*) Building height (*b*) Building line

   (*c*) Covered area (*d*) Floor area ratio.
10. Briefly explain various requirement of drawing to be submitted for getting permission for constructing a building.

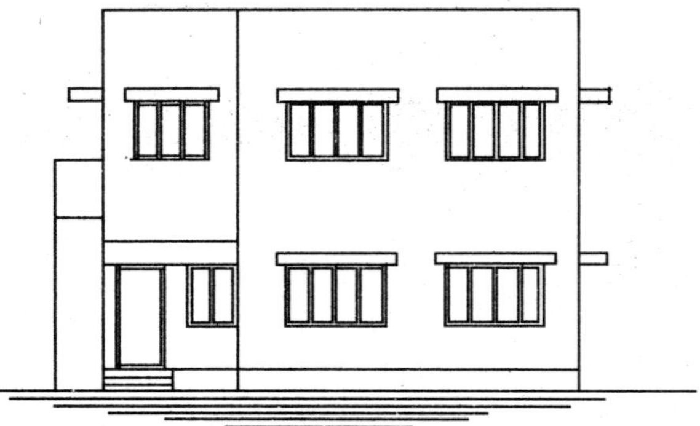

FRONT ELEVATION

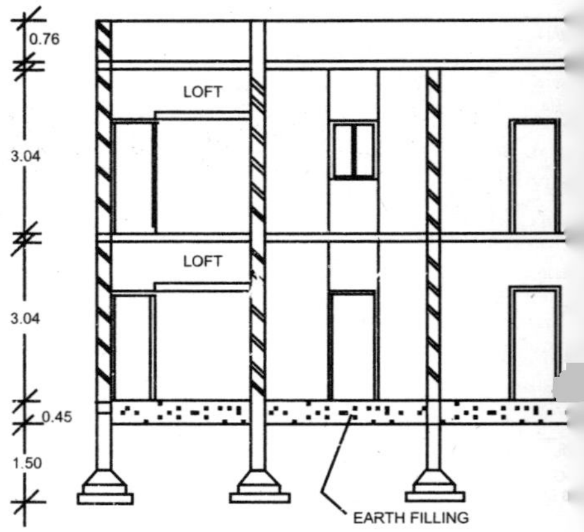

SECTION AT 'A' 'A'

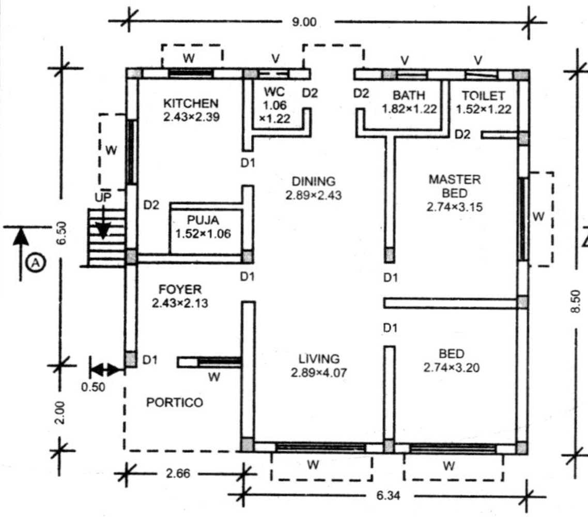

GROUND FLOOR PLAN

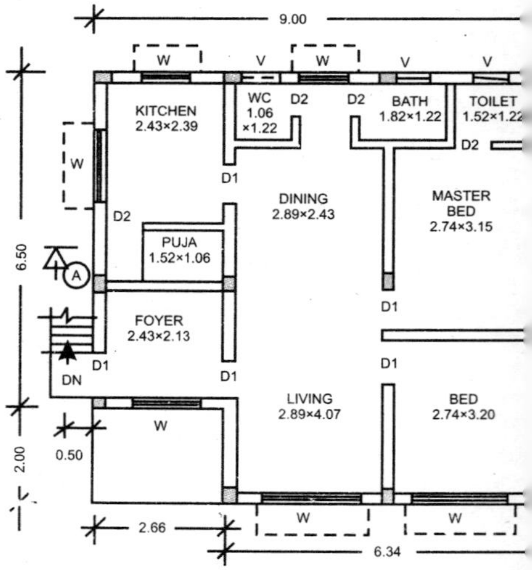

FIRST FLOOR PLAN

SCALE = 1:100
DRG. NO. =
DATE =

PROPOSED RESIDENTIAL BUILDING FOR MR. L
R.S. NO./BL NO. 100, 101, 102, 105, 106, 107, 108
92, 93, 96, 97, 81 : 2, 82 : 1 & 78 AT SATTUR HUD

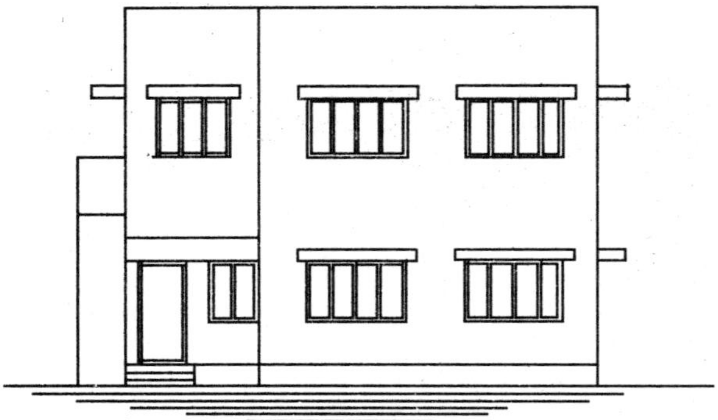

FRONT ELEVATION

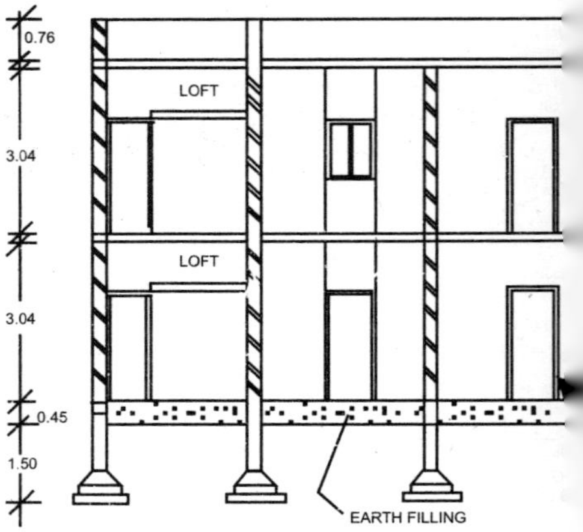

SECTION AT 'A' 'A'

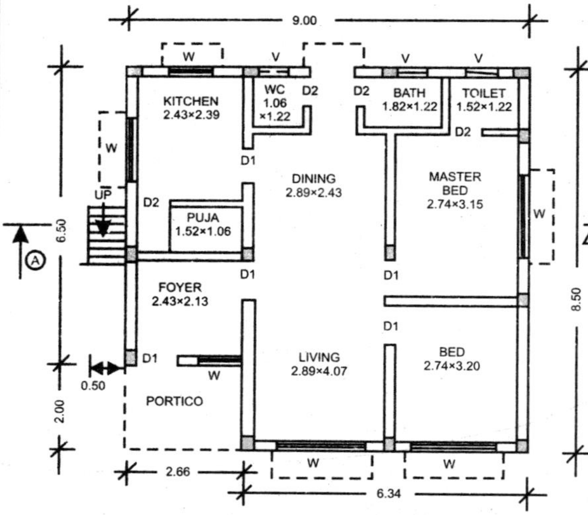

GROUND FLOOR PLAN

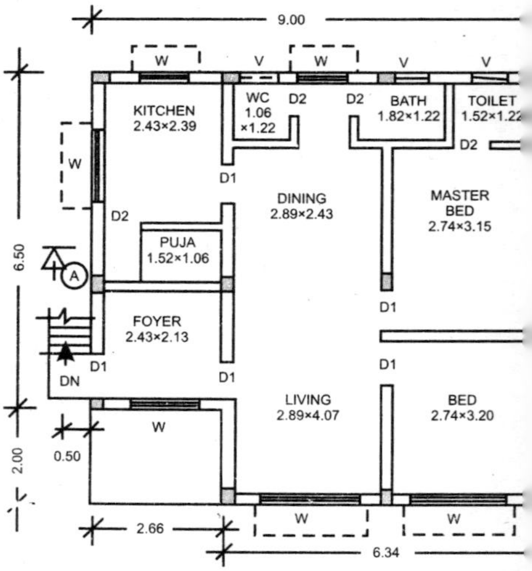

FIRST FLOOR PLAN

SCALE = 1:100
DRG. NO. =
DATE =

PROPOSED RESIDENTIAL BUILDING FOR MR. L
R.S. NO./BL NO. 100, 101, 102, 105, 106, 107, 108
92, 93, 96, 97, 81 : 2, 82 : 1 & 78 AT SATTUR HUD

ತಮ್ಮ ನಿವೇಶನದಲ್ಲಿ ಮಳೆ ನೀರಿನ
ಸಂಗ್ರಹಣೆ ವ್ಯವಸ್ಥೆಯನ್ನು
ಅಳವಡಿಸಿಕೊಳ್ಳತಕ್ಕದ್ದು

ಪ್ರಮಾಣ ಪತ್ರ
ಜೊತೆಗಿನ ಅಂಗೀಕೃತ ಯೋಜನೆ (ನಕ್ಷೆ)

ನಗರ ಯೋಜಕ ಸದಸ್ಯರು,
ಹುಬ್ಬಳ್ಳಿ ಧಾರವಾಡ ನಗರಾಭಿವೃದ್ಧಿ ಪ್ರಾಧಿಕಾರ,
ಹುಬ್ಬಳ್ಳಿ-25

ಕಮೀಶನರ
ಹುಬ್ಬಳ್ಳಿ ಧಾರವಾಡ ನಗರಾಭಿವೃದ್ಧಿ ಪ್ರಾಧಿಕಾರ
ಹುಬ್ಬಳ್ಳಿ

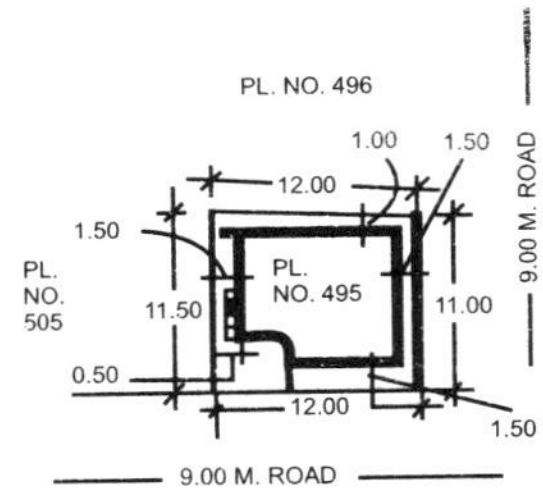

SITE PLAN
(SCALE = 1:500)

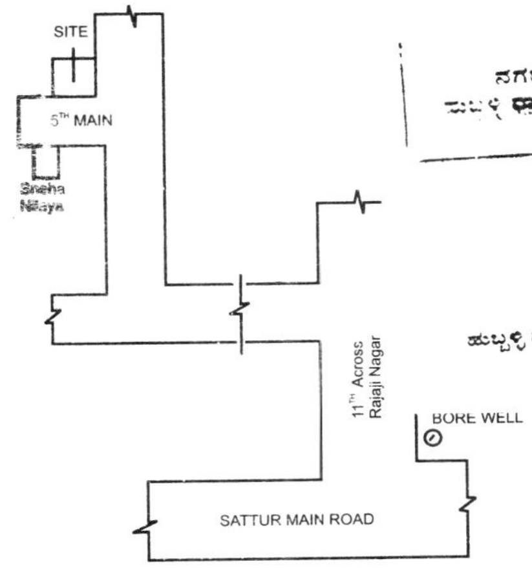

LOCATION MAP
(NOT TO SCALE)

AREA STATEMENT

PLOT AREA =	135.00 SQM
PROPOSED G.F. BUILT UP AREA =	71.18 SQM
PROPOSED F.F. BUILT UP AREA =	71.18 SQM
TOTAL BUILT UP AREA =	142.36 SQM
F. A. R. =	1.0 S
COVERAGE =	52.73%
COMPOUND WALL LENGTH =	23.00 RMTS

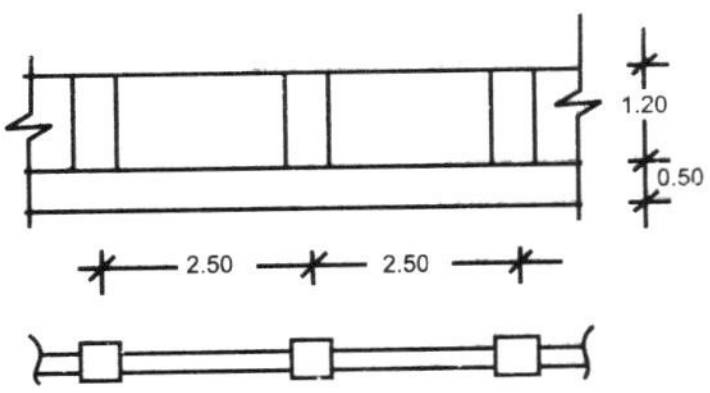

COMPOUND WALL

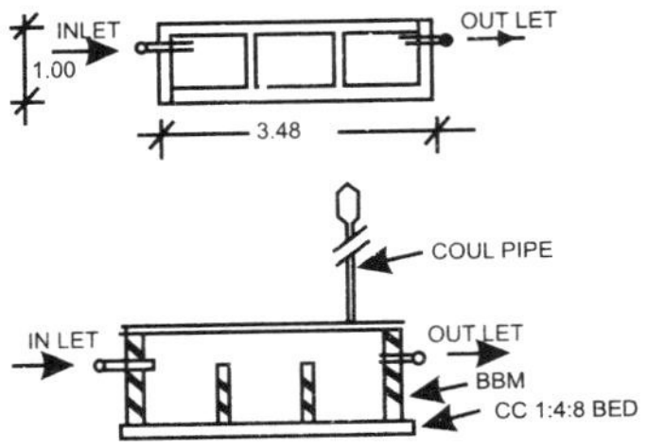

SEPTIC TANK DETAILS

NOTE :

(1) ALL DIMENSIONS ARE SHOWN IN METERS

Sd/—
APPLICANT SIGNATURE

STABILITY CERTIFICATE

THIS IS TO CERTIFY THAT THE FOUNDATIONS ARE DESIGNED STRONG ENOUGH TO TAKE THE LOAD OF GROUND & FIRST FLOOR. THE OVERALL STRUCTURE WILL BE SOUND AND SAFE

Sd/—

ENGINEER

SANJAY PATIL
CONSULTING ENGINEER
SHRI SHAKI, KEMPAGERI,
LINE BAZAR, DHARWAD
DHARWAD-580001

10

Water Resources Development

No life is possible without water. Hence it is necessary to utilise water judiciously and conserve it so that it is available at all time. In this chapter elementary hydrology is presented.

Water requirement for various purposes is discussed and different modes of storage structures, conveyance systems are briefly given.

10.1 ELEMENTARY HYDROLOGY

Hydrology may be defined as the science that deals with the processing, governing the depletion and replenishment of water resources of the land areas on the earth.

97.3% of water on earth is in oceans. Most of the remaining water is in the solid form near north and south pole. Only 1×10^{-5} of the total water takes part in hydrological cycle.

Hydrological cycle is a continuous process of exchange of water from the surface of the earth and ocean to the atmosphere and vice versa, solar heat playing major role in evaporating water and putting it into atmosphere while rain plays major role to put it back on the earth. The total hydrological cycle is divided into the following three zones:

1. Atmosphere—Gaseous form above hydrosphere.
2. Hydrosphere—Bodies of water on earth.
3. Lithosphere—Water below hydrosphere (underground water).

Water from hydrosphere evaporates due to heat from sun and forms part of atmosphere. The water vapour forms into cloud and go up to a height of 20 km. Then it gets condensed and rains. The rain water partly gets into ground and moves slowly while surface water flows in the form of streams and rivers to join ocean. Part of it may get evaporated also. This water cycle activity extends up to 16 km in atmosphere and to a depth of 800 m below ground surface. This water in hydrological cycle is the area of interest in hydrology.

10.2 SOURCES OF WATER

Though major source of water is the ocean and that near the north and south pole in the form of ice, the 1×10^{-5} part of water that is involved in the hydrological cycle is main source of hydrology. Rain is the major source of useful water . It is stored in the tanks. The water that flows as streams and rivers is next important source of hydrology. By building bunds and dams it is stored in reservoirs. Underground water forms water table. It is tapped for use by digging open wells or bore wells. The average annual precipitation falling over the earth is about 5,77,000 billion cubic metres, of which 80% is on ocean and 20% is on the land.

10.3 WATERSHED DEVELOPMENT

For any cross-section of a stream or river, the area from which rain water flows in is called its water shed. This can be easily found from contour maps. The topo sheets help in indentifying the extent of area. The sites at which small lengths of bunds or dams help to impound large volume of water in reservoir are selected for constructing bunds and dams.

India is getting rains in mansoon period. Hence it needs lot of storage of water for other period, when streams and rivers start drying. A proper water shed management is very much essential to meet the requirement of water for entire year. Hence ever since India became independent, lot of importance has been given to construction of dams and canals. Still lot of work is to be done to meet water need for drinking and irrigation. Table 10.1 shows average annual availability and surface storage of various important river basins in India.

Table 10.1 Water resources potential and present surface storage of basins of India

Sr. No.	*River basin*	*Average annual availability in Bm³*	*Present surface storage in Bm³*
1.	Indus	73.3	16.3
2.	Ganga	525.0	40.3
3.	Brahmputra	585.6	2.4
4.	Godavari	110.5	24.3
5.	Krishna	78.1	42.6
6.	Mahi	11.0	4.9
7.	Sabarmati	3.8	1.4
8.	Narmada	45.6	10.3
9.	Tapi	14.9	10.2
10.	Cauvery	21.4	12.5

India has constructed about 2000 large dams storing about 220 Bm^3 of water. But still it is only 30% of the contemplated capacity. Still an area of 100 mega hectare is drought-prone and 40 mega hectare is flood prone. To balance it, there is proposal to connect Ganga-Cauvery rivers. So far the construction of reservoirs were in ideal places, submerging mainly government lands. Differences among the states about sharing water is widening and it may pose greater problems for water shed management in future. To co-ordinate watershed management a National Water Resources Council has been set up by Government of India, so that schemes can be implemented smoothly and quickly.

10.4 WATER REQUIREMENT

Water is generally required for the following purposes:

1. For municipal and industrial water supply
2. For irrigation
3. For hydropower generation
4. For navigation.

1. **Municipal and Industrial Demand:** Water is required for drinking, cooking, bathing, gardening and sanitary uses. On an average, the domestic consumption of water in India is 135 litres per day per person.

   Industrial water supply includes water supply to offices, hotels, hostels, hospitals and factories. On an average this demand is 20 to 25% of the total public water consumption.

   Apart from these, the water is required for public uses, airports, railways for fire fighting etc. Hence another 5% may be added to the total demand of the city. In general, a municipal area needs 270 litres/day/person.

2. **Water Requirement for Irrigation:** Every crop needs certain quantity of water for its growth. Total quantity of water required for the full-fledged growth is expressed in terms of centimetre depth standing over the irrigated area is called delta of that crop. This water is required over a suitable intervals of the growth of crop. This requirement may be met by rain also.

   The average value of delta is the highest for sugarcane and paddy which is 120 cm. For wheat it is only 30 cm. Maize and barly need only 25 to 30 cm delta. There is loss of water in carrying water from a reservoir to field due to evaporation and seepage. This should be accounted while finding the area that can be irrigated from a reservoir.

3. **Water for Hydropower Generation:** Hydroelectric power generation is the cheapest mode of power generation, particularly in long run. It needs storage of water to see that required quantity of water is let in power stations throughout the year. A certain minimum head of water is required for this purpose. The water used for power generation can be further used for other purposes in downstream. Hence practically there is no wastage of water in hydroelectric power generation.

4. **Water for Navigation:** Navigation provides cheap mode of transport and hence contributes for the economical development of the area. Actually it will not consume any water but minimum depth of water is to be maintained for the safe navigation. This depth is 2.7 m. A depth of 3.7 m is preferred. However in the process of maintaining minimum depth a certain quantity of water flows into sea.

10.5 CONSERVATION OF WATER

Water is a precious material, its importance is felt by many when water supply is disrupted. It should be conserved so that it is available in the lean period also. The various means of conservation may be classified as:

1. Minimising Transpiration Loss
2. Suppressing Evaporation Losses
3. Avoiding Seepage Loss
4. Avoiding Wastage
5. Recharging Underground Storages.

1. **Minimizing Transpiration Loss:** Transpiration is the process by which water vapour escapes from the living plant, mainly through the leaves and enters the atmosphere. The ratio of water transpired to water used by the plant is as high as 800. Type of plant, soil and climate influence transpiration. Evapotranspiration loss may be reduced by the following measures:

(*i*) By destroying the unwanted plant that transpire efficiently.

(*ii*) By growing plant varies that transpire less.

(*iii*) By reducing air movement over the crop by planting rows of taller trees.

2. **Suppressing Evaporation Losses:** Evaporation losses from reservoirs can be suppressed to conserve the water. The methods of suppressing evaporation losses may be grouped into the following three:

(*i*) Reducing surface area

(*ii*) Providing mechanical covers

(*iii*) Providing surface films.

(*i*) *Reducing surface area:* The volume of water lost due to evaporation is directly proportional to exposed surface area of reservoirs. Hence evaporation can be reduced by reducing exposed surface area. Some of the possible means are

(*a*) Select reservoir site such that ratio of area to storage is least; in other words, deeper reservoirs are to be preferred.

(*b*) If possible store water below ground level.

(*c*) Store water in large reservoirs instead of going for large number of small reservoirs.

(*d*) Prefer straight channels.

(*ii*) *Providing mechanical covers:* These methods are possible for smaller reservoirs. The types of mechanical covers to be provided include roofs, floating rafts, floating wax blocks, floating light weight concrete blocks, etc.

(*iii*) *Providing surface films:* Polar compounds act as evaporation suppressors. Cetyl alcohol is the commonly used polar compound. It forms a thin film, which is not visible for naked eye and reduces evaporation loss. About 25 gram of it is sufficient to form a micro layer over one hectare of water surface. Some investigators have claimed the reduction of 33% evaporation losses by using these barriers.

However the wind birds and insects disturb this layer and reduce its effectiveness. Hence nowadays research is in using a plastic net over these films.

3. **Avoiding Seepage Loss:** There is considerable seepage loss in canals. This may be reduced by providing impervious lining to canals or by carrying water to field by conduit canals.

4. **Avoiding Wastage:** Wastage of water in domestic use, public places and irrigation is common in India. Because of intermitent water supply, many times taps are not closed and when there is water supply, lot of water is simply wasted. Public should be aware of avoiding wastage of water. They should inform the concerned authorities if some pipelines are broken. Breaching and poor maintenance of canals result into wastage of water. Ill informed farmers use unnecessarily high quantity of water for crops. They should be trained to use optimum quantity. Drip irrigation and sprinkler irrigation can conserve lot of water.

In India water is taxed least. The amount collected is not sufficient even to maintain the system. Metering of water supply and avoiding subsidising water will certainly help to great extent in conserving water.

5. **Recharging Underground Storages:** This is the best way of storing water since it is cost effective and evaporation losses are avoided. The following three methods are generally adopted for recharging ground water:

(*i*) Spreading method (*ii*) Recharge well method

(*iii*) Induced infiltration method.

(*i*) *Spreading method:* In this method rain water is stored in pits or openland from where it directly infiltrates into shallow aquifiers.

(*ii*) *Recharge well method:* In the area with low permeability soil, spreading method is not effective. In such cases water is used to recharge open wells and bore wells. However water is to be purified before using it to recharge wells.

(*iii*) *Induced infiltration method:* In this method wells are constructed near the rivers. The percolating water is collected in the well and is stored in the lower level aquifier as shown in Fig. 10.1.

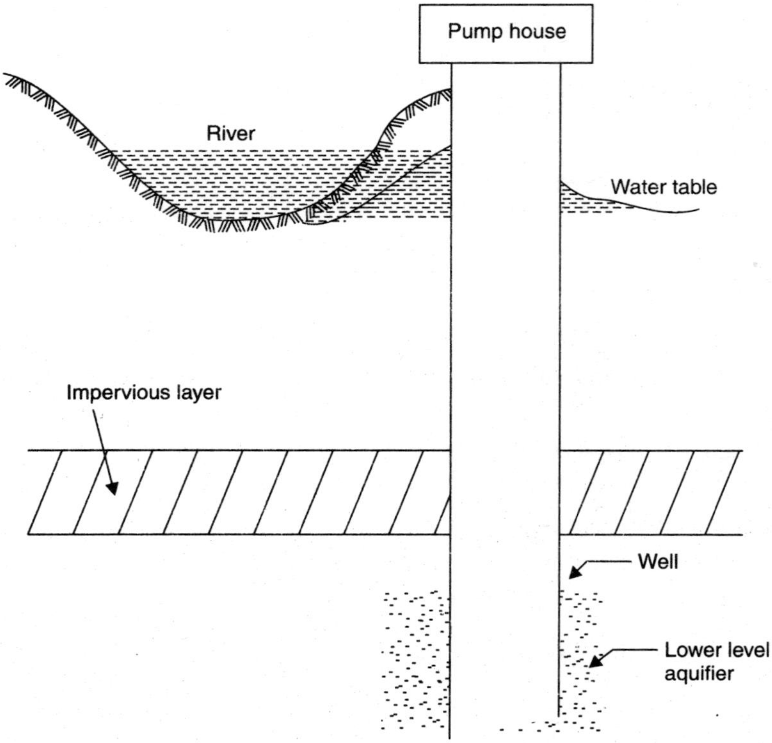

Fig. 10.1

10.6 HYDRAULIC STRUCTURES OF STORAGE

Bunds and dams are the hydraulic structures built to store water.

10.6.1 Bunds

Bunds are built across streams to impound water and make tanks. They are for minor irrigation or for drinking water facility of small towns and for villages. The height of bund may be of the order of 3 to 5 metres. To create foundation for bunds loose soil on the site is stripped and sandy layers removed. Sandy soil is not used for making bunds, since they are highly pervious. Gritty and gravelly soil may be used but clay is preferable. However to keep construction cost low, available earth from tank site is used for constructing bunds. Figure 10.2 shows two typical cross-sections of earthen bunds. Care should be taken to avoid failure of bunds due to

(*i*) Percolation

(*ii*) Slipping of slopes

(*iii*) Topping or overflow

(*iv*) Erosion by waves.

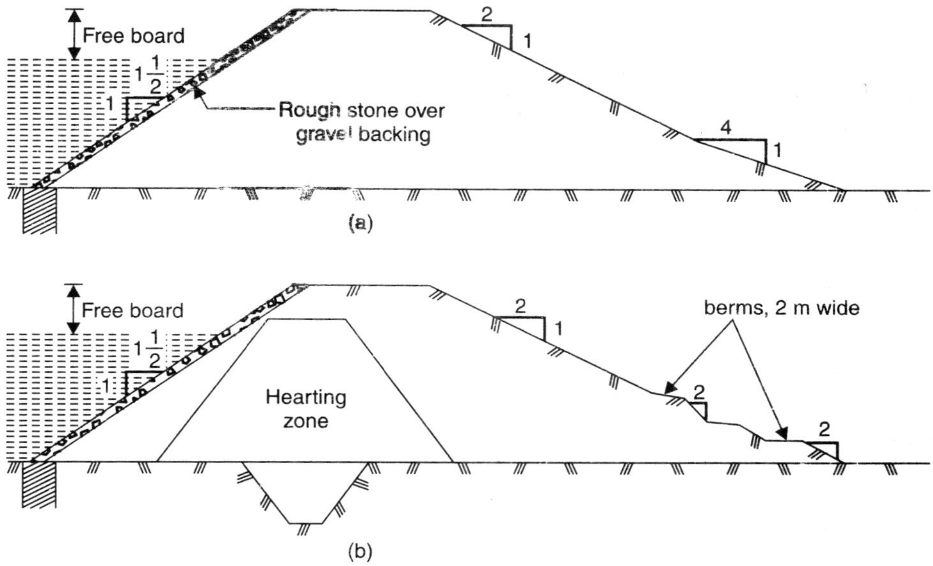

Fig. 10.2 Earthen bunds

10.6.2 Dams

Dams are the storage structures of greater heights built across rivers to impound large quantity of water and develop larger reservoirs. Nurek dam in Russia is over 300 m high and it is the highest dam in the world. Some of the high dams in India are Bhakra dam (226 m), Tehri dam (250 m) and Iddukki dam (170.7 m). Dams may be classified according to

(*i*) Materials used (*ii*) Structural design.

According to materials used dams may be classified as *earth dams, rockfill dams, masonry dams* and *concrete dams.* According to structural designs dams are classified as *gravity dams* and *arch dams.* Gravity dams resist water pressure by their weight. When the resultant of horizontal pressure of water retained and self weight of dam, pass through middle third of dam, the dam is safe. Tension will

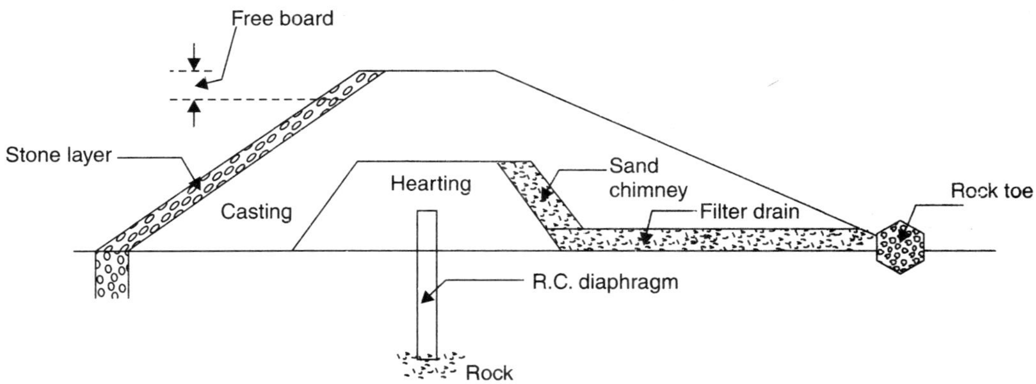

Fig. 10.3 Earth dam

not develop at toe of dam. Arch dams are having curved profile in plan and they resist water pressure by arch action, *i.e.,* as compressive force in the arch with slight bending. There may be a single arch or multiple arches to resist the water pressures. Arches need strong abutments to resist heavy thrust. All earth dams, rockfill dams and masonry dams are gravity dams, while concrete is used for the construction of arch dams. Idduki dam (Kerala) is an arch dam. Figures 10.3 to 10.7 show typical cross-sections of various types of dams:

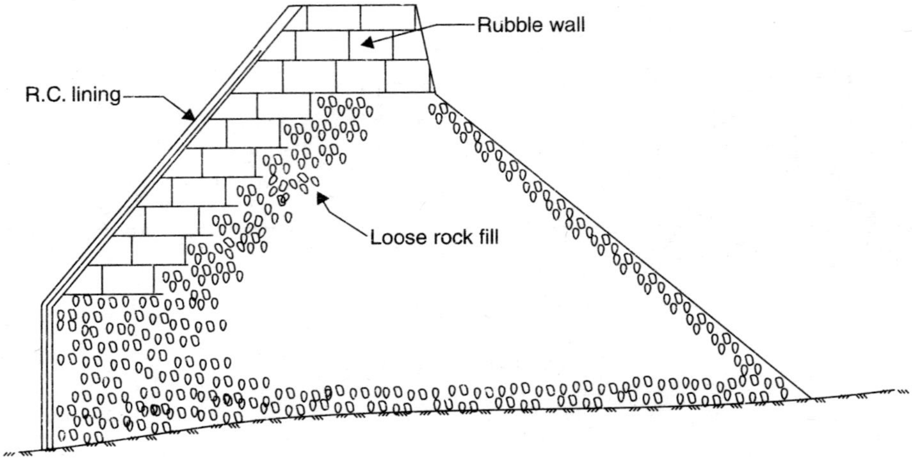

Fig. 10.4 Rockfill dam

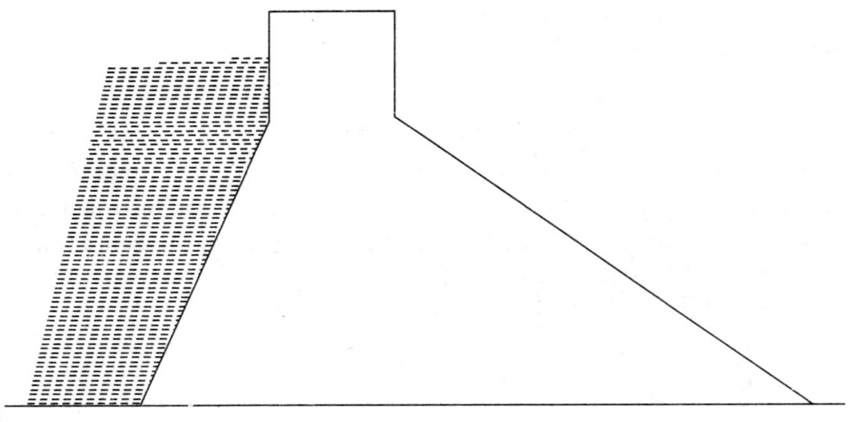

Fig. 10.5 Masonry dam

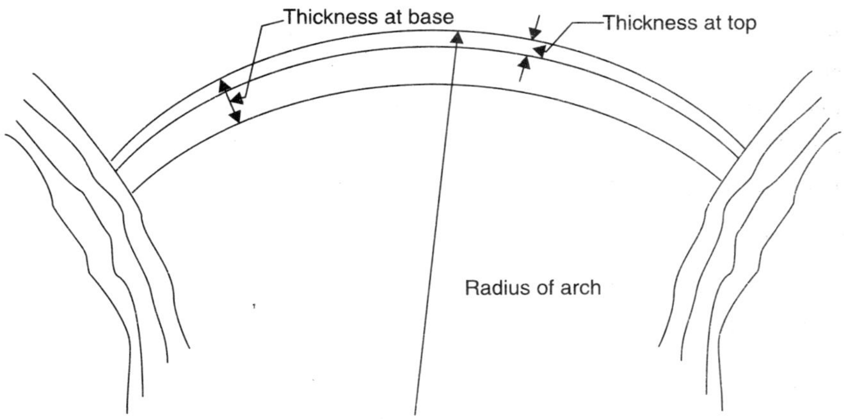

Fig. 10.6 Arch dam (Plan view)

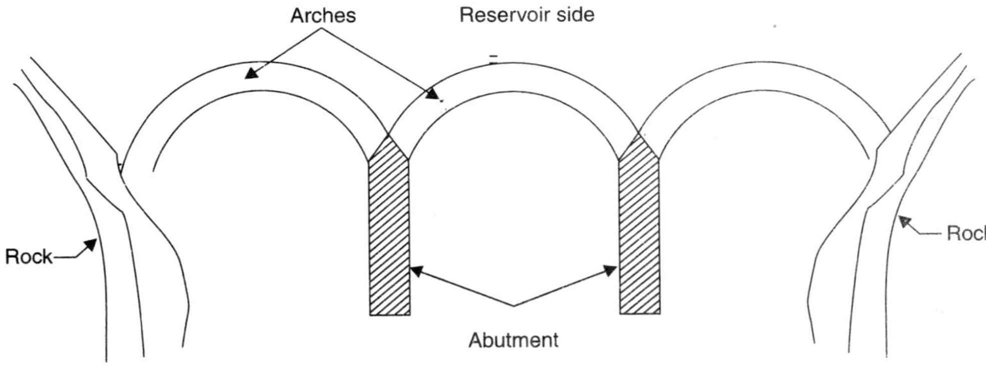

Fig. 10.7 Multiple arch dam (Plan)

10.7 WATER CONVEYANCE SYSTEM

Canals and conduits are the water conveyance systems, which are commonly used.

10.7.1 Canals

Canals are the artificial channels dug for carrying water from a river or reservoir to the needy places. In general they have trapezoidal sections. According to the purpose they serve, they are designated as irrigation canals, power canals and navigational canals. In this article we are basically concerned about irrigation canals. These canals are classified on the basis of various consideration as presented below:

(*i*) *On the basis of nature of source of supply:* On this basis irrigation canals are classified as permanent canals and inundation canals. Permanent canals get water from a reservoir or from a perennial river. They are provided with permanent regulation and distribution works. Inundation canals get water from rivers which are not perennial. Only when level of river rises above a limit, water starts flowing in these canals. For this purpose simple cuts are provided in the river bank. They do not have specially designed diversion or regulatory works.

(*ii*) *On the basis of function of canal:* On this basis canals are classified as feeder canal and carrier canals. Feeder canal carries water from the source only to feed to another canal. They never irrigate fields directly. Carrier canals may feed other distributaries but will feed agricultural field also.

(*iii*) *On the basis of importance in a network:* On this basis canals may be classified as

(*a*) Main canal (*b*) Branch canal

(*c*) Major distributary (*d*) Minor distributary

(*e*) Field channel.

(*iv*) *On the basis of canal alignment:* On this basis irrigation canals are classified as

(*a*) Ridge canals (*b*) Contour canals

(*c*) Side slope canals.

Ridge canals run along the ridge lines topology of the area. Hence they can irrigate field on both sides. Since they are on the ridge, they do not need cross drainage works.

Contour canals are almost along the contour lines and hence can serve fields at far away places also. They can irrigate field on only one side.

Side slope canals are almost at right angles to the contour lines. They can serve fields on both sides. But water head is lost in a short distance.

(*v*) *Based on lining being provided:* On this basis canals may be classified as lined canals and unlined canals. Canals are provided with impervious beds and sides to prevent seepage. Usually main canals and branch canals are lined. Distributary canals are not lined, to save cost of lining.

(*vi*) *On the basis of soil through which canal is constructed:* On this basis canals may be classified as alluvial canals and non-alluvial canals. Alluvial canals are through alluvial soil while non-alluvial canals are through rocky area. In Indo-Ganges planes most of the canals are alluvial canals.

(*vii*) *Based on financial output:* On this basis irrigation canals are classified as productive canals and protective canals. Productive canals when completed yield net revenue not less than maintenance cost plus 6 per cent interest on the investments. Protective canals do not yield much income but they give relief to famine affected areas.

10.7.2 Conduits

Conduits of suitable size can replace the canals. However, they are costly. At present conduits are used mainly to transport water from the nearby bore wells for irrigating fields. PUC conduits are very popular conduits. However for municipal water supplies, concrete coated steel pipes are used.

QUESTIONS

1. Explain the term hydrological cycle.
2. Write short notes on:
   (*a*) Sources of water
   (*b*) Watershed development
   (*c*) Water requirement.
3. Write explanatory note on conservation of water.
4. Give a typical cross-section of earthen bund and state various causes for their failures.
5. Give typical cross-section of the following types of dams:
   (*a*) Earth dam
   (*b*) Rock fill dam
   (*c*) Masonry dam
   (*d*) Arch dam
   (*e*) Multiple arch dam.
5. Write short notes on canals.

11

Transportation Engineering

Transportation is the most important facility in moving men and material and it has considerable role in economical progress. Planning, building and maintaining roads and railways is a major part of transportation engineering. It also includes airways and waterways facility to be provided. For proper planning of these transportation facility, traffic survey is to be carried out. Managing traffic efficiently in available facility is also part of transportation engineering.

In this chapter role of transportation in national development is presented. Types of transportations are listed. Elements of highway material properties and highway construction is explained. Elements of traffic engineering and traffic control is presented.

11.1 ROLE OF TRANSPORTATION IN NATIONAL DEVELOPMENT

Transportation facility plays very important role in nation developments. These benefits can be listed below:

1. **Economical Growth:** Connecting producing centres to marketing centres avoids lot of unnecessary wastes. Imports and exports become easy. By this both producer and consumers are benefitted, resulting into economical growth. More industries will come up to cater to various needs of the public.
2. **Social Effects:** Due to economical development activities, man power is utilised for the benefit of mankind. Exploitation of rural people is minimised, since the former can avoid middlemen and directly supply his produces to marketing places. More frequent travels by people make them broad minded and reduce sectionalism. National and international integration improves. Antisocial activities come down.
3. **Safety and Law and Order:** Efficient transportation of law enforcing forces bring safety to the public. The security of the nation improves with better transport facility.
4. **Disaster Management:** In case of distress due to floods, earthquake, etc. the suffering of the affected people can be minimised by rushing aid, if good transportation facility exists.

Thus transportation facility plays very important role in economical development and prosperity and peace of the nation.

11.2 TRANSPORTATION WAYS

The following are the common methods of transportation

1. Roadways
2. Railways
3. Airways and
4. Waterways

Apart from these four, the following types of transportations are required in specific cases:

1. Conveyors
2. Pipelines and
3. Ropeways.

11.3 SURFACE TRANSPORTATION AND AIRWAYS

11.3.1 Surface Transportation

Roadways and railways are the popular modes of transportation on the surface of land and hence they are jointly known as surface transportation.

Development of civilisation is closely associated with the developments of roads. The first hard surface was constructed in 3500 BC in Mesopotamia. Almost at the same time good roads were built in Harappa and Mohenjo-Daro. In 600 BC metallic road of 6 to 7.5 m wide existed in Rajgir near Patna. In about 300 BC, Kautilya got National Highway constructed connecting North-West Frontier Province (Now in Pakistan) and Patna. Shershah got a highway constructed connecting Lahore and Sonargaon (Bengal). All the rulers who contributed to construction of roads are very well remember in the history.

Various criteria may be used for classifying roads. Depending upon the usage of roads during rainy season, they may be classified as

(*i*) all weather roads and
(*ii*) fair weather roads.

All weather roads are not flooded during rainy season. In fair weather roads overflowing of streams across the road is permitted during peak monsoon season.

Based on the type of pavement surfaces provided, the roads may be classified as:

(*i*) surfaced roads
(*ii*) unsurfaced roads.

Surfaced roads are provided with a finishing surface of bitumen or cement concrete while unsurfaced roads may be mud roads or water bound mechadam layer road.

Road construction and maintenance is a costly affair. Hence width and quality of road should be suitably selected depending upon the type and volume of traffic. Based on this, in India, the following classification is made:

(*i*) Express Highways
(*ii*) National Highways (NH)
(*iii*) State Highways (SH)
(*iv*) Major District Roads (MDR)
(*v*) Minor or Other District Roads (ODR) and
(*vi*) Village Roads (VR,.

(*i*) **Express Highways:** These are most superior highways and they are built wherever volume of traffic is very high. They have superior facilities and design speed. No cross traffic is permitted on express highways. They are provided with central separator to separate traffic in opposite directions. Controlled access is provided to other roads and towns. They are fenced to keep away animals. For catering to local traffic separate feeder roads are provided adjoining express highways. Only fast moving vehicles are permitted on express highways. Golden Quadrilateral connecting Delhi, Mumbai, Chennai, Kolkata and Delhi is first express highway work taken up in India.

(*ii*) **National Highways (NH):** These are the roads connecting important cities, towns, ports etc. of different states. They may even connect neighboring countries also. The National Highways have two lane traffic, at least 8 m wide and with at least 2 m wide shoulders on each side. The construction and maintenance of these roads is taken care by the central government agencies like CPWD or military engineering services (MES). The National Highways are assigned the numbers. The highway connecting Delhi–Ambala–Amritsar is NH–1.

(*iii*) **State Highways:** These are the roads built by state governments to connect important cities of the state. Central government may give grants for building and maintaining these roads. These highways also have 8 m carriage way and 2 m wide shoulders on each side.

(*iv*) **Major District Roads (MDR):** These are roads within a district connecting markets and production areas to State or National highways or railway stations. The MDR has lower speed and geometric design specifications compared to those for NH or SH.

(*v*) **Minor or Other District Roads (ODR):** These roads connect rural areas of production to market centres, taluka centres or other main roads. These roads have lower design specifications than MDR. These roads are looked after by district authorities with the help of state government departments.

(*vi*) **Village Roads:** The roads connect a villages or a groups of villages to the roads of higher category. The local district boards are responsible for the construction and maintenance of these roads. These roads are usually unmetalled.

For long distance travel and transportation of goods, railway is advantageous, because they workout cheaper. It is only 1/4th to 1/6th of that required by road.

Railways are classified as narrow gauge, metre gauge and broad gauge. Nowadays it is felt that unigauge is preferable. Hence wherever possible nowadays all railways are being converted to broad gauge (1.68 m gauge).

11.3.2 Airways

For the long distance movement, airways is the fastest. It needs construction and maintenance of airports. The disadvantages of airways is expensive and it is occasionally disturbed by weather conditions.

11.4 ELEMENTS OF HIGHWAY MATERIAL PROPERTIES

The cross-section of carriage way consists of the following components [Fig. 11.1]:

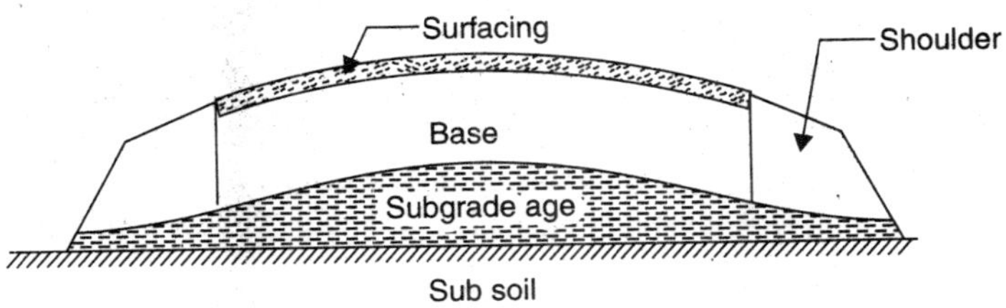

Fig. 11.1 Cross-section of a carries way

(*a*) Subsoil (*b*) Subgrade

(*c*) Base and (*d*) Surfacing.

(*a*) **Subsoil** is the natural or prepared soil which will take the load of the road. It is prepared by properly compacting the natural soil.

(*b*) **Subgrade** gives support to the road structure. It should remain stable and dry throughout. Considerable attention should be given to laying proper subgrade to get stable road surface. It is made of gravel, sand, silt and clay.

(*c*) **The base** may consists of two layers, top layer being called as base and bottom layer as sub base. Sub base is made with stabilised soil or selected granular soil, bricks or boulders. Base course is provided with broken stone aggregates.

(*d*) **Surfacing** is the topmost layer of carriage way which takes loads from traffic directly. It provides a smooth nonslippery and stable surface for the vehicles. It should be impervious enough to protect subgrade from rainwater. It may be provided with cement concrete or with bituminous concrete.

Thus, the materials required for the construction of road are

(*a*) Soil (*b*) Aggregate

(*c*) Bitumin/tar and (*d*) Cement.

The above materials selected should possess the required properties like

(*a*) Stability (*b*) Incompressibility

(*c*) Permanency of strength under adverse conditions also

(*d*) Good drainage and (*e*) Ease of compaction

The required properties of each material is discussed and necessary tests are briefly listed in this article.

11.4.1 Soil Properties and Tests

Natural soil can be classified as sand, loam, silt and clay. Sand is the soil in which individual grains are seen. When dry it flows freely. Loam is the uniform mixture of sand, silt and day. Silt is like soft flour like soil, when it is dry. Clay forms hard lumps when dry. Actual soil consist of different mixture of above and also with organic materials. For example sandy clay, which has clay as main constituent but has a small quantity of sand.

In order to identify and classify soils, the following index properties are generally found:

(*i*) Grain size distribution

(*ii*) Attenberg's limits like – Liquid limit, plastic limit and shrinkage limit.

(*i*) **Grain Size Analysis:** It consists of sieve analysis for coarse grained particles and sedimentation analysis for soil fines. The sieve analysis consists in sieving a measured quantity of soil through standard sieves of smaller size and expressing weight retained on each sieve as percentage of total weight of sample. To find grain size distribution of fine soil that passes through 75 micron sieve, the principle that larger grains in a suspension settle faster is used.

(*ii*) **Attenberg's Limits:** Behaviour of soil with different moisture content is used for classifying clay. This type of tests are known as attenberg limit test. There are three different limit tests recommended. They are:

Liquid Limit

Plastic Limit and

Shrinkage Limit

Liquid Limit: The moisture content which causes soil to change from viscous to plastic stage is known as liquid limit of soil. It is defined as water content expressed as a percentage weight of the over dry soil, at the boundary between liquid and plastic states of consistency of soil. Clay is having higher liquid limit whereas sand and silts have lower values.

Plastic Limit: It is the percentage of water at which the soil changes from plastic to semi-solid state. It is expressed as a percentage of the weight of over dry soil when soil changes from plastic to semi solid state. It is found by finding least quantity of water the soil can be rolled into thread of 3 mm diameter without breaking into pieces. Sands and peats do not possess this property hence they have no plastic limit.

Shrinkage Limit: When water evaporates from soil its volume reduces up to a certain limit and after that no shrinking takes place. Hence shrinkage limit may be defined as the moisture content, below which reduction in volume will not take place.

Generally soils having high values of liquid limit and plastic limit are poor engineering soils.

Strength Properties of Soil

To evaluate strength properties of soil, the following three types of tests are conducted:

(*i*) Shear tests

(*ii*) Bearing tests

(*iii*) Penetration tests

Shear Tests: There are three methods of finding shear strength of soil as shown below:

(*a*) Direct shear test

(*b*) Triaxial compression test

(*c*) Unconfined compression test.

Direct Shear Test: It is carried out in a apparatus which consists of a box divided horizontally in two halves. The box is filled up with soil. A vertical load is applied and a horizontal pull is caused to produce a specified rate of displacement of movable part of the box. The horizontal force resisted is indicated by dial gauge. The maximum horizontal force which cause shearing of soil is noted. Test is carried with different vertical pressure. Then normal stress (vertical stress) vs. shear stress curve as shown in Fig. 11.2 is plotted. The values of cohesion c and angle of internal friction ϕ are found.

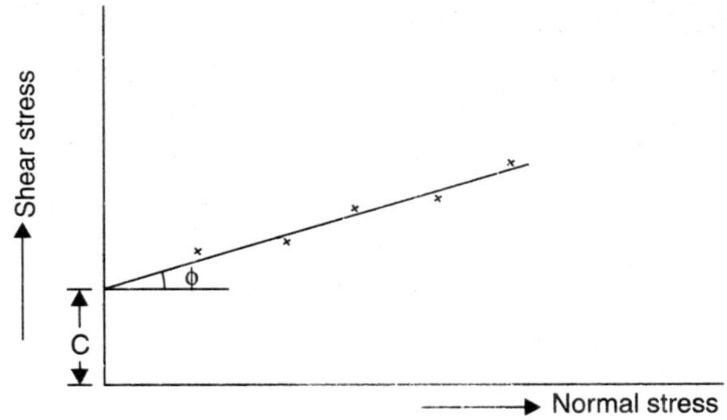

Fig. 11.2 Direct shear test results

Triaxial Compression Test: In actual field soil is confined by surrounding soil and is subjected to vertical load. This situation is simulated in the test known as triaxial test. A cylindrical specimen is prepared and inserted in a thin membrane, placed in a triaxial cell. Triaxial pressure is then applied to specimen. The vertical load is gradually increased till failure load is reached. Usually a constant strain rate of 1.25 mm per minute is applied. The specimen fail either by shearing or by bulging. The tests are carried out at different lateral pressures preferably 0, 0.75 and 1.5 kg/cm^2. The values of confining pressure and vertical pressure are plotted and semi circle is drawn passing through these points. The common tangent to Mohr's rupture envelope gives C and ϕ and hence shearing stress can be found (Fig. 11.3).

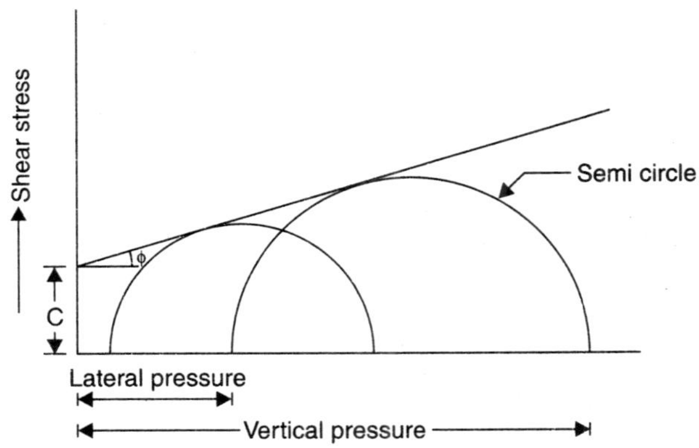

Fig. 11.3 Triaxial test results

Unconfixed Compression Test: It is nothing but a compression test with zero lateral pressure. Hence a simple machine like compression machine can be used. Strain is applied at a rate of 1/2 to 2% per minute. Maximum vertical load is taken corresponding to 20% deformation or till cracks appear, whichever is less.

Bearing Test

The plate load test is conducted to find load bearing capacity of the pavement or soil. In this test a set the plates of diameters 750, 600, 450 and 300 mm are placed one over the other on the surface to be tested. A vertical load is gradually increased recording the settlement of plates from an undisturbed frame. The load corresponding to yielding of surface is noted. Figure 11.4 shows a typical arrangement for such a test.

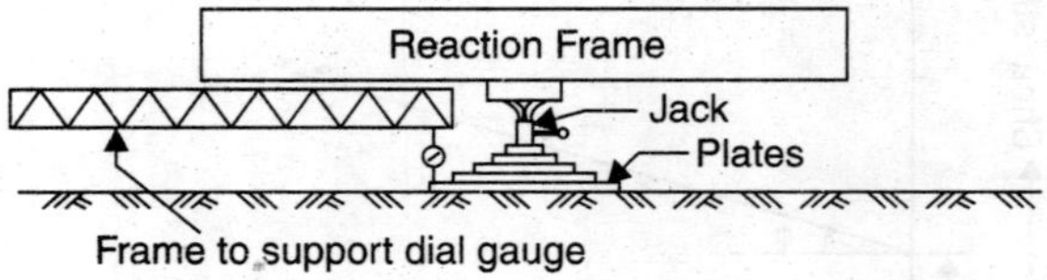

Fig. 11.4 Plate bearing test arrangement

California Bearing Ratio (CBR) Test

This is a penetration test. In this the strength of subgrade and bases of roads are found by determining the penetration resistance of the soil. This test is conducted in the laboratory on a prepared specimen in a mould or in situ in the field.

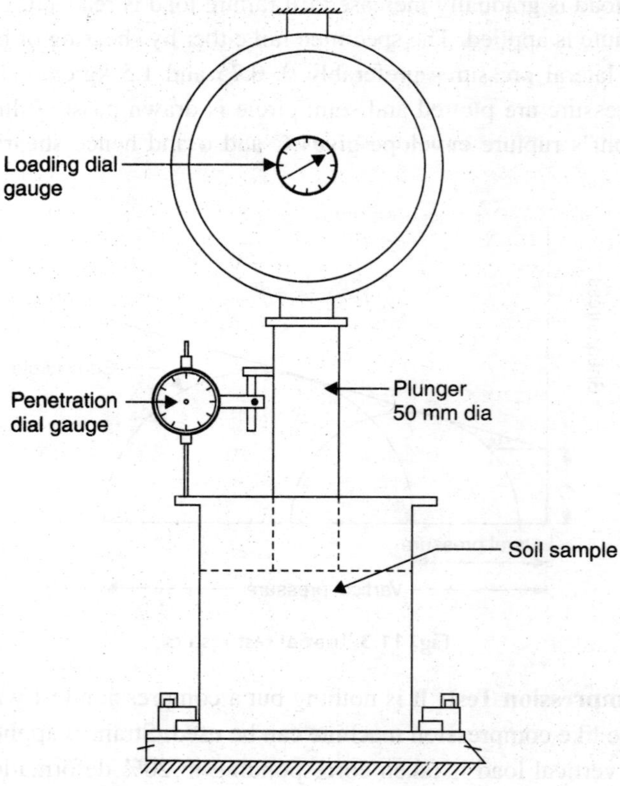

Fig. 11.5 C.B.R. test

11.4 ELEMENTS OF HIGHWAY MATERIAL PROPERTIES

The cross-section of carriage way consists of the following components [Fig. 11.1]:

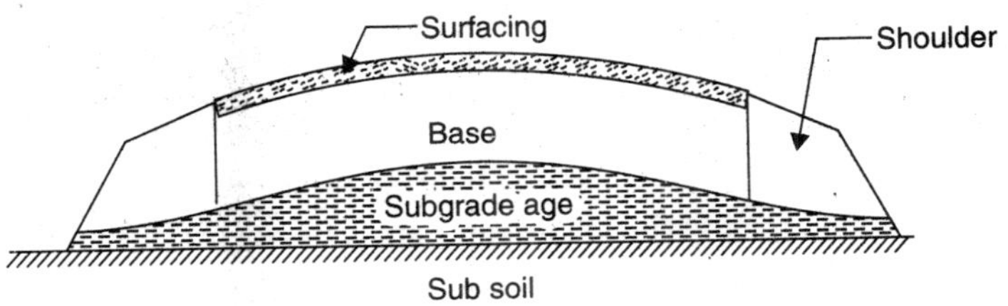

Fig. 11.1 Cross-section of a carries way

(*a*) Subsoil (*b*) Subgrade
(*c*) Base and (*d*) Surfacing.

(*a*) **Subsoil** is the natural or prepared soil which will take the load of the road. It is prepared by properly compacting the natural soil.

(*b*) **Subgrade** gives support to the road structure. It should remain stable and dry throughout. Considerable attention should be given to laying proper subgrade to get stable road surface. It is made of gravel, sand, silt and clay.

(*c*) **The base** may consists of two layers, top layer being called as base and bottom layer as sub base. Sub base is made with stabilised soil or selected granular soil, bricks or boulders. Base course is provided with broken stone aggregates.

(*d*) **Surfacing** is the topmost layer of carriage way which takes loads from traffic directly. It provides a smooth nonslippery and stable surface for the vehicles. It should be impervious enough to protect subgrade from rainwater. It may be provided with cement concrete or with bituminous concrete.

Thus, the materials required for the construction of road are

(*a*) Soil (*b*) Aggregate
(*c*) Bitumin/tar and (*d*) Cement.

The above materials selected should possess the required properties like

(*a*) Stability (*b*) Incompressibility
(*c*) Permanency of strength under adverse conditions also
(*d*) Good drainage and (*e*) Ease of compaction

The required properties of each material is discussed and necessary tests are briefly listed in this article.

11.4.1 Soil Properties and Tests

Natural soil can be classified as sand, loam, silt and clay. Sand is the soil in which individual grains are seen. When dry it flows freely. Loam is the uniform mixture of sand, silt and day. Silt is like soft flour like soil, when it is dry. Clay forms hard lumps when dry. Actual soil consist of different mixture of above and also with organic materials. For example sandy clay, which has clay as main constituent but has a small quantity of sand.

In order to identify and classify soils, the following index properties are generally found:

(*i*) Grain size distribution

(*ii*) Attenberg's limits like – Liquid limit, plastic limit and shrinkage limit.

(*i*) **Grain Size Analysis:** It consists of sieve analysis for course grained particles and sedimentation analysis for soil fines. The sieve analysis consists in sieving a measured quantity of soil through standard sieves of smaller size and expressing weight retained on each sieve as percentage of total weight of sample. To find grain size distribution of fine soil that passes through 75 micron sieve, the principle that larger grains in a suspension settle faster is used.

(*ii*) **Attenberg's Limits:** Behaviour of soil with different moisture content is used for classifying clay. This type of tests are known as attenberg limit test. There are three different limit tests recommended. They are:

Liquid Limit

Plastic Limit and

Shrinkage Limit

Liquid Limit: The moisture content which causes soil to change from viscous to plastic stage is known as liquid limit of soil. It is defined as water content expressed as a percentage weight of the over dry soil, at the boundary between liquid and plastic states of consistency of soil. Clay is having higher liquid limit whereas sand and silts have lower values.

Plastic Limit: It is the percentage of water at which the soil changes from plastic to semi-solid state. It is expressed as a percentage of the weight of over dry soil when soil changes from plastic to semi solid state. It is found by finding least quantity of water the soil can be rolled into thread of 3 mm diameter without breaking into pieces. Sands and peats do not possess this property hence they have no plastic limit.

Shrinkage Limit: When water evaporates from soil its volume reduces up to a certain limit and after that no shrinking takes place. Hence shrinkage limit may be defined as the moisture content, below which reduction in volume will not take place.

Generally soils having high values of liquid limit and plastic limit are poor engineering soils.

Strength Properties of Soil

To evaluate strength properties of soil, the following three types of tests are conducted:

(*i*) Shear tests

(*ii*) Bearing tests

(*iii*) Penetration tests

Shear Tests: There are three methods of finding shear strength of soil as shown below:

(*a*) Direct shear test

(*b*) Triaxial compression test

(*c*) Unconfined compression test.

Direct Shear Test: It is carried out in a apparatus which consists of a box divided horizontally in two halves. The box is filled up with soil. A vertical load is applied and a horizontal pull is caused to produce a specified rate of displacement of movable part of the box. The horizontal force resisted is indicated by dial gauge. The maximum horizontal force which cause shearing of soil is noted. Test is carried with different vertical pressure. Then normal stress (vertical stress) vs. shear stress curve as shown in Fig. 11.2 is plotted. The values of cohesion c and angle of internal friction ϕ are found.

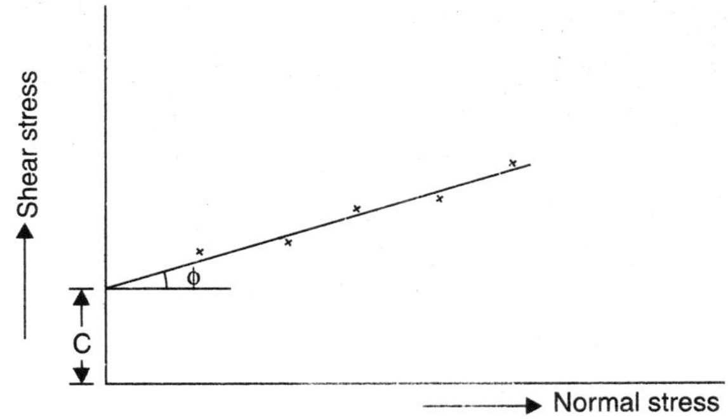

Fig. 11.2 Direct shear test results

Triaxial Compression Test: In actual field soil is confined by surrounding soil and is subjected to vertical load. This situation is simulated in the test known as triaxial test. A cylindrical specimen is prepared and inserted in a thin membrane, placed in a triaxial cell. Triaxial pressure is then applied to specimen. The vertical load is gradually increased till failure load is reached. Usually a constant strain rate of 1.25 mm per minute is applied. The specimen fail either by shearing or by bulging. The tests are carried out at different lateral pressures preferably 0, 0.75 and 1.5 kg/cm^2. The values of confining pressure and vertical pressure are plotted and semi circle is drawn passing through these points. The common tangent to Mohr's rupture envelope gives C and ϕ and hence shearing stress can be found (Fig. 11.3).

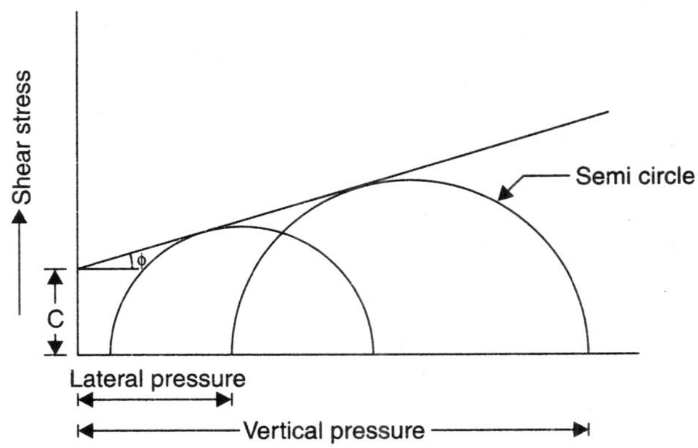

Fig. 11.3 Triaxial test results

Unconfixed Compression Test: It is nothing but a compression test with zero lateral pressure. Hence a simple machine like compression machine can be used. Strain is applied at a rate of 1/2 to 2% per minute. Maximum vertical load is taken corresponding to 20% deformation or till cracks appear, whichever is less.

Bearing Test

The plate load test is conducted to find load bearing capacity of the pavement or soil. In this test a set the plates of diameters 750, 600, 450 and 300 mm are placed one over the other on the surface to be tested. A vertical load is gradually increased recording the settlement of plates from an undisturbed frame. The load corresponding to yielding of surface is noted. Figure 11.4 shows a typical arrangement for such a test.

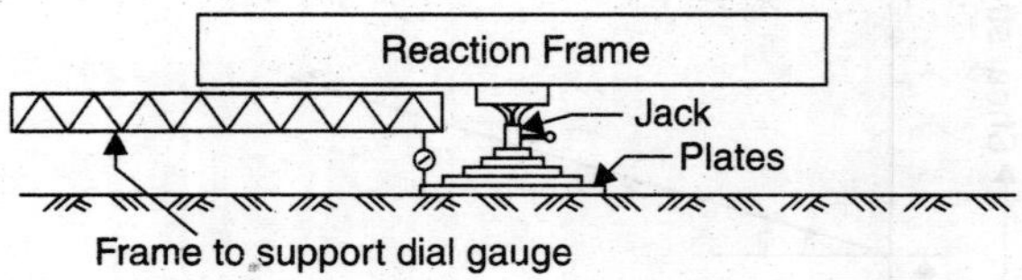

Fig. 11.4 Plate bearing test arrangement

California Bearing Ratio (CBR) Test

This is a penetration test. In this the strength of subgrade and bases of roads are found by determining the penetration resistance of the soil. This test is conducted in the laboratory on a prepared specimen in a mould or in situ in the field.

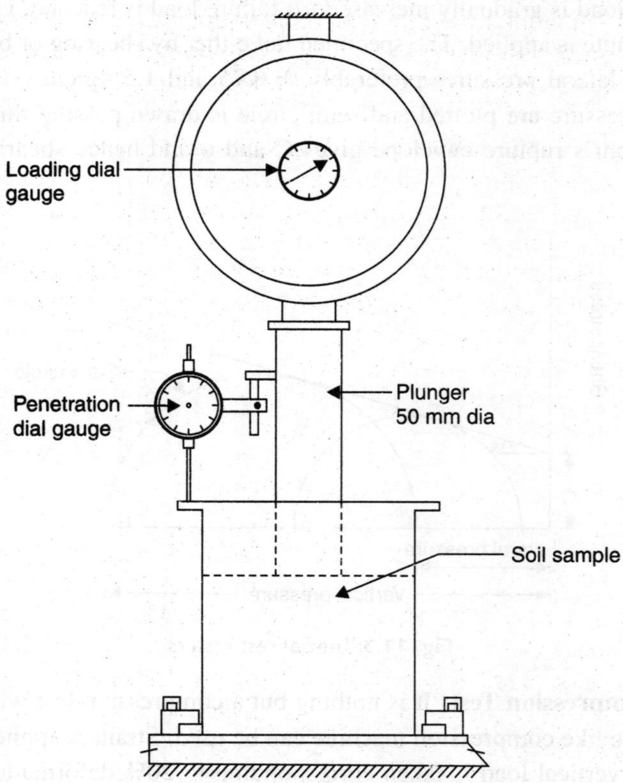

Fig. 11.5 C.B.R. test

Figure 11.5 shows a typical CBR test apparatus used in the laboratories. It consists of a 50 mm diameter plunger to penetrate and a dial gauge to measure the amount of penetration. The load is applied by means of a screw jack or hydraulic jack at a uniform rate of penetration of 1.25 mm/minute. The load is applied on a remoulded soil sample and recorded at penetration of 0, 0.5, 1.0, 1.5, 2.0, 2.5, 3.0, 4.0, 5.0, 7.5, 10.0 and 12.5 mm. The load penetration curve is plotted. From the graph load corresponding to 2.5 mm and 5.0 mm are found.

$$\text{Then CBR value} = \frac{P_T}{P_S} \times 100$$

where P_T – Corrected test load obtained from graph

P_S – Standard load.

Standard load for 2.5 mm is 13440 N and for 5.0 mm is 20258 mm. The CBR value is found for both penetration and average is taken as CBR value.

The test procedure should be strictly followed. It is an arbitrary strength test giving idea about the strength but not actual value of strength.

11.4.2 Properties and Tests on Road Aggregates

The properties of an ideal aggregate for use in road construction are as follows:

(*i*) It should be hard (*ii*) It should be strong

(*iii*) It should be tough (*iv*) It should be durable

(*v*) It should have low moisture absorption (*vi*) It's specific gravity should be high

(*vii*) It should have proper shape so as to interlock properly when compacted

Granite and basalt suit as road materials. The following tests are specifically recommended to check the suitability of aggregates as road material:

(*i*) Crushing test (*ii*) Abrasion test

(*iii*) Impact test

Crushing Test: The testing apparatus consists of a 152 mm diameter cylinder and a plunger of size 115 mm diameter and height 180 mm. The aggregate is filled in 3 equal layers, each layer tamped 25 times by the tamper. Then a load of 400 kN is applied at 40 kN/minute through the plunger. The aggregate is removed from the cylinder and sieved on 2.36 mm size I.S. Sieve.

$$\text{The aggregate crushing value is} = \frac{W_2}{W_1} \times 100$$

where W_2 – Weight of material passing through sieve

and W_1 – Original weight of sieve.

Lower the aggregate crushing value, stronger is the aggregate. The crushing value should not be more than 45% for base course and 30% for surface course. Figure 11.6 shows aggregate crushing apparatus.

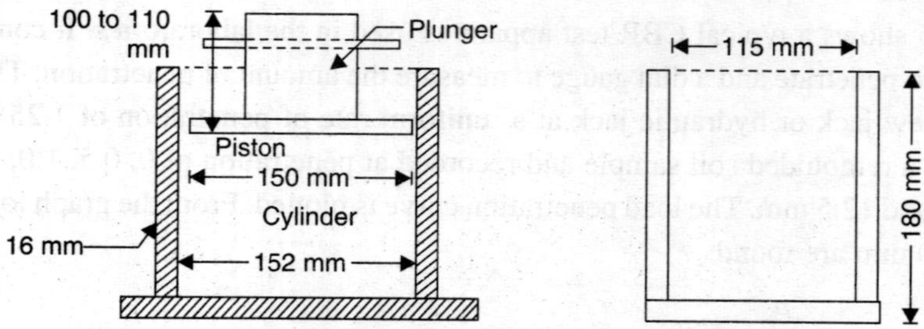

Fig. 11.6 Aggregate crushing apparatus

Abrasion Test: The aggregates used in the surface coat should be hard enough to resist the abrasion due to traffic. Any one of the following three tests may be conducted to find the suitability of aggregates:

(*i*) Los Angeles abrasion test. (*ii*) Deval abrasion test

(*iii*) Dorry abrasion test.

However Los Angeles abrasion test is preferred as the test results are having good correlation with the performance of pavements. This test is briefly explained below:

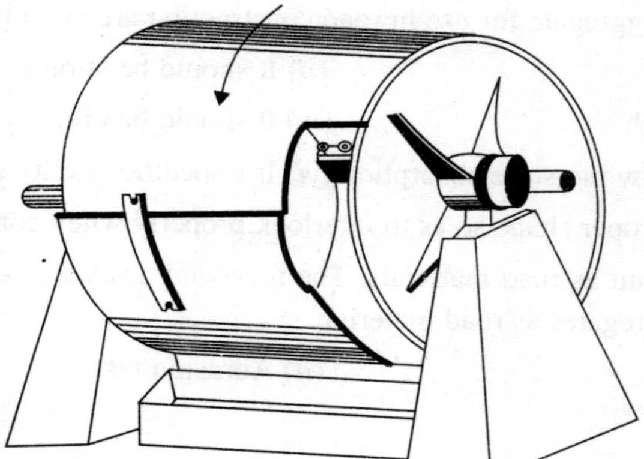

Fig. 11.7 Los Angeles abrasion testing machine

The Los Angeles apparatus [Fig. 11.7] consist of a hallow cylinder 0.7 m inside diameter and 0.5 m long with both ends closed. It is mounted on a frame so that it can be rotated about horizontal axis. IS code has standardised the test procedure for different gradation of specimen. Along with specified weight of specimen specified number of cast iron balls of 48 mm diameter are placed in the cylinder. The cylinder is rotated at a speed of 30 to 33 rpm for specified number of times (500 to 1000). Then aggregate is removed and sieved on 1.7 mm IS sieve. The weight of aggregate passing is found. Then Los Angeles value is found as

$$= \frac{\text{Weight of aggregate passing through sieve}}{\text{Original weight}} \times 100$$

The following abrasion values are recommended for road works:

For bituminous mixes -- 30%

For base course -- 50%

Impact test: The resistance of stone to impact loads is found as explained below:

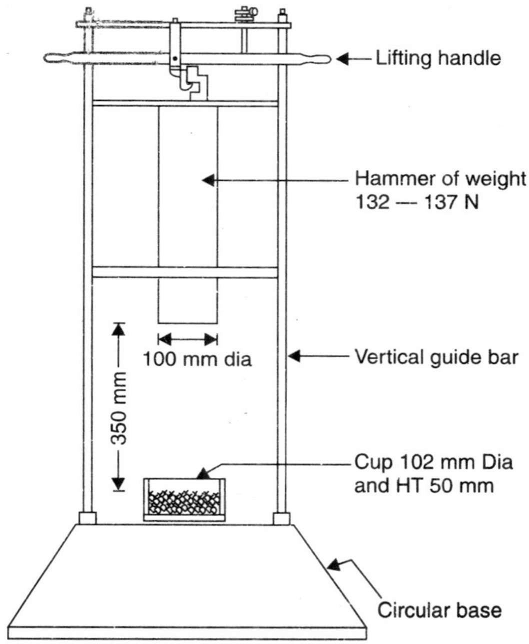

Fig. 11.8 Aggregate impact testing machine

Figure 11.8 shows line diagram of impact testing machine. It consists of a frame with guides in which a metal hammer of weight 13.5 to 15 kg can freely fall from a height of 380 mm.

Aggregates of size 10 mm to 12.5 mm are filled in cylinder in 3 equal layers, each layer being tamped 25 times. The same is then transferred to the cup and again tamped 25 times. The hammer is then allowed to fall freely on the specimen 15 times. The specimen is then sieved through 2.36 mm sieve. Then

$$\text{Impact value} = \frac{W_2}{W_1}$$

where W_2 = weight of fines

and W_1 = original weight.

The recommended impact values for various works are:

(*i*) For wearing course $\not>$ 30%

(*ii*) For bituminous mechadam $\not>$ 35%

(*iii*) For water bound mechadam $\not>$ 40%.

11.4.3 Properties of Bituminous Materials

Bitumen and tar are the bituminous materials used for the construction of roads. They should possess the following properties:

(*i*) Viscosity should be adequate at the time of mixing. For this they should be heated to suitable temperature.

(*ii*) There should be adequate adhesion between bitumen and aggregate.

(*iii*) During summer it should not become too soft and during winter too hard resulting into cracking.

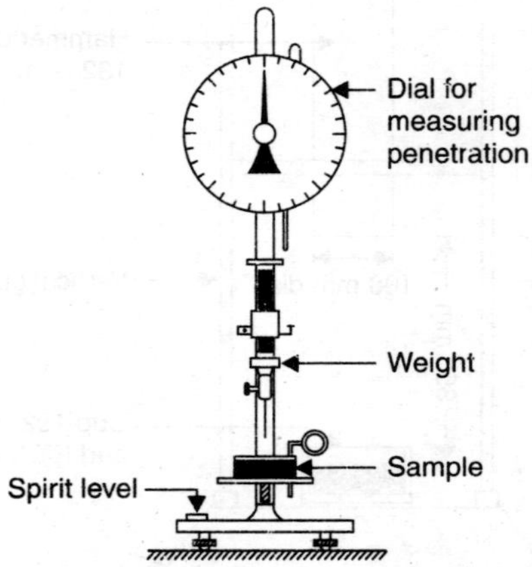

Fig. 11.9 Penetrometer test

Penetration test is conducted to determine hardness/softness of bitumen. The test is conducted in measuring the depth of penetration of standard needle under standard load in 5 seconds [Fig. 11.9]. Penetration value is specified as *a*/*b* which means it is *a* to *b* divisions. 1 division of penetration means 0.1 mm. The common grades adopted in India are 30/40 and 80/100 depending upon climatic conditions. Figure 11.9 shows the penetrometer used for testing.

The ductility is another important property to ensure proper binding between bitumen and aggregates. The ductility is expressed as the distance in centimetres to which a standard briquette of bitumen can be stretched before the thread breaks. Figure 11.10 shows test set up. The cross-section of test specimen at minimum width is 10 mm × 10 mm.

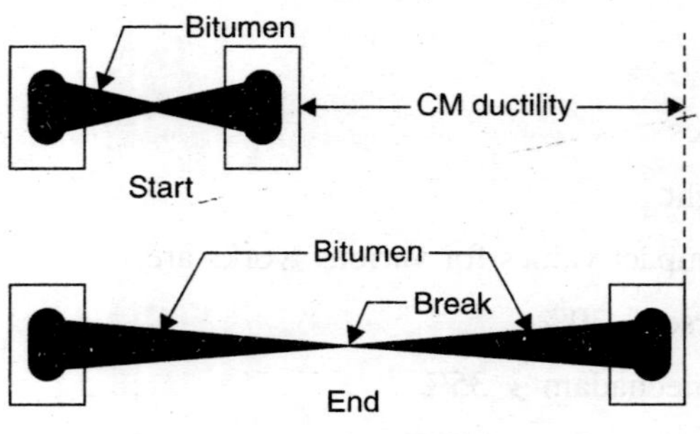

Fig. 11.10 Ductility test

A minimum value of 75 mm has been specified by Bureau of Indian Standards.

Many believe that viscosity test is better measure than penetration test to find suitability of bituminous materials. Viscosity is the measure of resistance to flow.

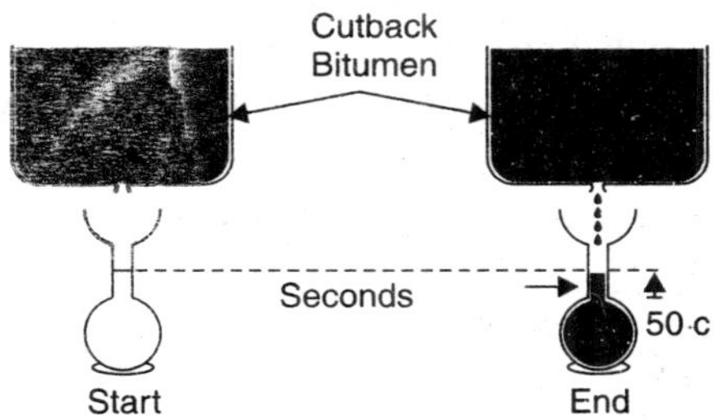

Fig. 11.11 Viscosity test set-up

The test conducted to measure viscosity is with the arrangement shown in Fig. 11.11. It is measured as time taken in seconds for 50 mL of tar to flow through 10 mm orifice of the standard viscometer at specified temperature. For bitumen sample it is through 4 mm orifice at 25°C or 10 mm orifice at 25 to 40°C.

There are five grades of tars RT-1, RT-2, RT-3, RT-4 and RT-5. RT-1 has low viscosity and RT-5 has high viscosity. Depending upon the climatic conditions tars of different grades are recommended for road works.

11.4.4 Cement

Cement is used in the construction of cement concrete pavements. The required properties of cement have been already presented in Art 8.4.

11.5 ELEMENTS OF HIGH WAY CONSTRUCTION

Construction of all roads are carried out in two phases:

1. Earth work and preparation of subgrade
2. Construction of pavement.

11.5.1 Earthwork and Preparation of Subgrade

According to the alignment and topography of the area the subgrade is prepared either in embankment or in cutting. After marking center line and alignment the construction activities are carried out:

Construction in Embankment: It involves the following:

(*i*) Clearing	(*ii*) Stripping
(*iii*) Compacting original ground	(*iv*) Placing embankment
(*v*) Rolling	(*vi*) Finishing

(*i*) ***Clearing:*** The obstruction such as trees, bushes and buildings are first cleared from the construction limits of the road.

(*ii*) ***Stripping:*** The top loose soil is stripped and stored aside to be used latter for embankment slopes to grow vegetation.

(*iii*) ***Compacting Original Ground:*** For a width of subgrade plus 1 m on either side original ground is compacted by rolling or tamping. Normally height of embankment should be 0.6 m. If it is not possible, original ground is loosened to 0.5 m and rolled in layers of 0.25 m.

(*iv*) ***Placing of Embankment:*** The suitable soil is brought from borrow pits and laid in layers of 250 mm over entire width. The moisture content is brought to 1 to 2% above optimum moisture content before rolling. The hard lumps are broken.

(*v*) ***Rolling:*** Each layer of soil in embankment should be thoroughly rolled to get required compaction. Depending upon the type of road and soil, following types of rollers are used:

Smooth wheeled roller

Pneumatic typed roller

Vibratory rollers

Sheep foot roller and

Power rammer

Each compacted layer should be tested for density achieved before placing the next layer. Shoulders are also constructed on the same way. The embankment is brought to the desired grade and camber and levels checked.

(*vi*) ***Finishing Operations:*** Embankments are finished according to the alignment, levels and camber. The top soil stored is spread over the slopes. Grass is planted on the slopes so as to minimise erosion.

Construction of Subgrade in Cutting

In cutting subgrade is brought to the desired grade and camber by cutting soil and removing it. Then the top is rolled.

11.5.2 Construction of Pavement

The following types of pavements are provided on roads depending upon the importance of the road.

(*i*) Earthen roads

(*ii*) Gravel roads

(*iii*) Soil stabilised roads

(*iv*) Water bound mechadam (WBM) roads

(*v*) bituminous black top roads

(*vi*) Cement concrete roads

Their construction is briefly explained below:

(*i*) ***Earthen Roads:*** It is a low cost road. The width of earthen road varies from 6 to 7.5 m. The thickness varies from 450 mm to 750 mm. Side drains are provided to carry rain water. They are constructed with soil available at site or near by. To drain out water from surface camber of 1 in 23 to 1 in 20 is provided. If camber is more than 1 in 20, there is danger of soil being eroded. Sufficient care should be taken to drain rain water.

(*ii*) ***Gravel Road:*** Compared to earthen road gravel roads can carry heavier traffic. The carriage way is made up of well compacted crushed rock or gravel. A camber of 1 in 25 to 1 in 30 is provided. The following two types of construction are employed

1. Trench type
2. Feather edge type.

The two types are shown in Fig. 11.12.

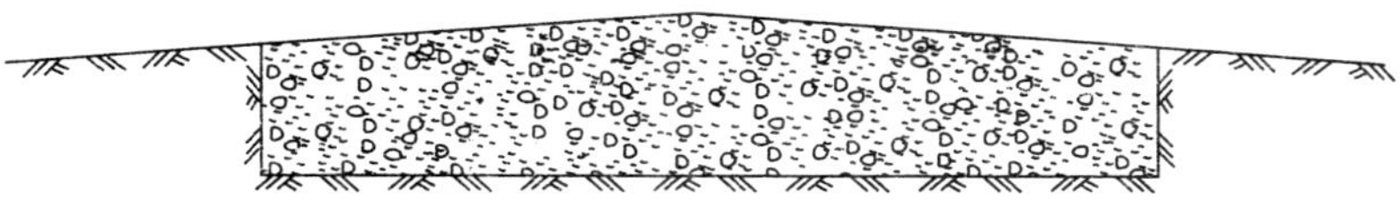

(a) Trench Type Gravel Road

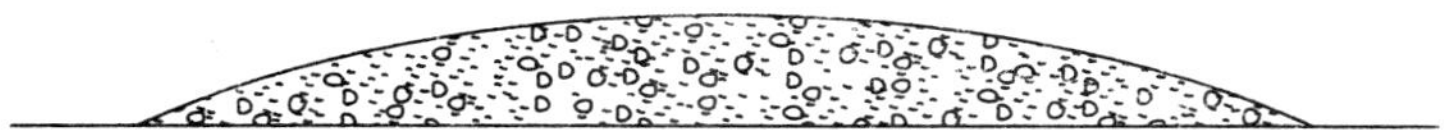

(b) Feather Edge Type Gravel Road

Fig. 11.12

The pavement is usually made in two layers over the sub grade. Lower layer consist of 100 mm thickness. The subgrade surface is scratched and this layer is placed and rolled by means of road roller. Water is sprinkled during compaction. The top layer is then placed and compacted to required thickness. Sheep foot roller or multi-tyred rollers are used for rolling.

(*iii*) ***Soil Stabilised Roads:*** Earthen roads may be upgraded by stabilising them. The following methods of stabilisations are possible

(*a*) Mechanical method – *i.e.,* by adding granular soil (sand)

(*b*) Stabilising with cement

(*c*) Soil lime stabilisation and

(*d*) Soil bitumen stabilisation.

After mixing stabilising materials top layer is compacted by rolling . When funds permit, stabilised pavements could form base course to bituminous or concrete roads.

(*iv*) ***Water Bound Mechadam Roads:*** It consists of clean coarsed aggregates which are interlocked by rolling and filling the voids in the aggregate by screening and binding materials. Camber of 1 in 36 to 1 in 48 is given depending upon the rainfall of the area.

Coarse aggregates used are:

Crushed or broken stone

Burnt broken bricks

Camber

Laterite.

The aggregates should be of graded sizes.

Screenings are used to fill up the voids between the coarse aggregates. Generally material is same as that of coarse aggregates. However screening is not required, if crushable type of aggregates such as brick are used.

Fine grained materials with plasticity index 4–9 are used as surfacing coarse.

Construction procedure for WBM consists of

(*a*) Preparation of foundation
(*b*) Provision of lateral confinement of aggregates
(*c*) Spreading coarse aggregates
(*d*) Rolling
(*e*) Providing screening
(*f*) Application of binding material.

(*a*) **Preparation of Foundation:** Subgrade should be prepared to required grade and camber.

(*b*) **Lateral Confinement of Aggregates:** This is done by constructing the shoulders first to the thickness of compacted layer. It prevents lateral displacement of aggregates when compacted.

(*c*) **Spreading Coarse Aggregates:** The coarse aggregate is placed and spread over the prepared surface in a layer of 75 mm thickness.

(*d*) **Rolling:** Usually they are rolled using 6 to 10 tonnes three wheel rollers. Slight sprinkling of water may be made during rolling. Rolling is stopped when the aggregates are sufficiently compacted to permit application of screening.

(*e*) **Screening:** Screening should not be dumped. It should be applied uniformly in thin layers and spread. It is followed by rolling and brooming layer by layer. Dry rolling is done to see that due to vibration screening material fills the voids. Sprinkling of water may be made and rolled at the end.

(*f*) **Application of Binding Material:** After application of screening binding material is applied at a uniform and slow rate in two or more thin layers. After applying each layer plenty of water is sprinkled over the surface. After sweeping with broom, surface is rolled with 6 to 10 tonne roller. During rolling water shall be applied to wheels of roller to wash down binding material stuck.

After final compaction of the road, 24 hours curing is allowed. Next day, wherever necessary screening and binding materials are applied and compacted. Traffic is allowed after the mechadam sets.

(*v*) ***Bituminous Black Top Roads:*** Bituminous black top roads are very commonly used, since they can be built in stages and upgraded whenever funds are available. The following two points should be carefully noted in bituminous black top constructions.

1. Viscosity of bituminous material changes with temperature and it is necessary to maintain correct viscosity during the construction.
2. Excess bituminous material is detrimental to the road surface.

Construction methods depend upon the types of bituminous surfaces adopted. The following types of bituminous surfaces are adopted:

(*a*) Interface treatment
(*b*) Surface dressing
(*c*) Penetration mechadam
(*d*) Premix bituminous pavements.

(*a*) ***Interface Treatment:*** Interface treatment is necessary to get binding between the existing surface and any other top layer to be provided. For this existing surface is cleaned to remove dust and dirt and a thin layer of bituminous binder is sprayed. If interface treatment is required on WBM base course, a binder of low viscosity is used so that it penetrates the surface. This is called *prime coat.* If the interface is required on a existing bituminous or concrete surface **track coat** is provided. Track coat is with bituminous material of higher viscosity.

(*b*) ***Surface Dressing:*** It is provided over an existing pavement to improve wearing surface. It may be in a single or in two layers. The first coat recommended is 12 mm thick and second coat 9 mm thick. Bitumen of grade 80/100 and 180/200 is first applied and then aggregate is uniformly spread. Then the surface is rolled using 6 to 8 tonne rollers.

(*c*) ***Penetration Mechadam:*** In this method, coarse aggregate is spread over a prepared base and packed to a uniform depth by rolling. Then bituminous material is applied by pressure distribution or by a nozzle attachment. The coarse aggregates may be provided on one or two layers. The thickness of each coarse should not exceed 75 mm. On top of it a seal coat is applied by first applying binder and then spreading chippings.

(*d*) ***Premix Bituminous Pavement:*** In these, aggregates and binders are mixed thoroughly and then it is spread over prepared base. In such construction each and every piece of aggregate is coated well with binder and may offer greater stability. The following types of premix bituminous pavements are provided:

- Bituminous bound mechadam
- Tar carpet
- Bituminous concrete
- Rolled asphalt and
- Mastic asphalt.

Bituminous Bound Mechadam: It consists of one or more coarses of crushed aggregates, premixed with bituminous binder and compacted. It serves as base coarse. It can be used as wearing pavement only after providing seal coarse.

Tar Carpet: After providing track coat to base, tar carpet of 20 mm thick premix layer is provided. Then it is finished with seal coat.

Bituminous Concrete: Bituminous concrete consists of bituminous material, 20 to 75 micron coarse aggregates and crushed screening or sand. Bitumen of grade 30/40, 60/70 or 80/100 are used for making this concrete. This concrete is prepared in a hot mix plant. The thickness of this concrete is usually 40 mm. For compacting 8–10 tonne rollers are used.

Rolled Asphalt: It is also known as sheet asphalt. It consists of bitumen, sand and filler but no coarse aggregates. It is used as wearing coat. It provides impervious surface.

Mastic Asphalt: It is also a mixture of bitumen, fine aggregates and fillers but cooked at temperature of 200 to 223°C. It is cooked in special cookers for 5 hours before spreading on the road surface. It is laid to a thickness of 25 to 50 mm. After cooling it forms hard surface which can absorb vibrations easily.

Hence it is commonly used as surface coarse to bridge deck slab.

(*vi*) ***Cement Concrete Roads:*** Cement concrete roads have excellent riding surface and are durable. However construction cost is high and are to be completed in a single stage. The construction involves the following stages:

(*a*) Preparing subgrade and sub-base
(*b*) Placing of forms
(*c*) Mixing of materials
(*d*) Transporting and placing of concrete
(*e*) Compacting and finishing
(*f*) Curing
(*g*) Removing of forms

(*a*) **Preparing Subgrade and Sub-base:** It should be at least 300 mm wider than concrete pavement to be provided. Minimum bearing capacity of soil to be achieved is 0.55 N/mm^2. The top 150 mm layer of soil should be compacted well.

(*b*) **Placing of Form:** Steel forms are preferred. They are in the forms of M.S. channel sections. The forms are cleaned and oiled before placing them on the subgrade tamped well. They are placed at 2 m interval. Usually continuous bay method of construction is preferred over alternate bay method [Fig. 11.13].

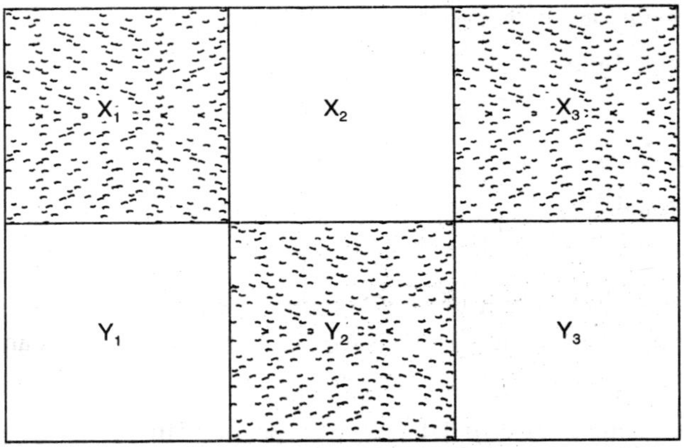

X_1, Y_2, X_3 - Alternate Bay Method
X_1, X_2, X_3 - Continuous Bay Method

Fig. 11.13

(*c*) **Mixing Materials:** Suitable concrete mix is designed to get proportion of cement, sand, coarse aggregate and water cement ratio so as to produce minimum flexural strength of 4 N/mm^2. In batching plant mixing of materials is carried out with correct proportion.

(*d*) **Transporting and Placing of Concrete:** Concrete is transported in wheel borrows or as head load in pans. Care should be taken to avoid segregation. It should be dropped on formation surface as close to it as possible. Spreading with showels should be least. Nearly 20% extra thickness is placed so that after compaction required thickness is obtained.

(*e*) **Compacting and Finishing:** The surface of pavement is compacted either by means of a power driven finishing machine or by vibrating. At corners and edges hand consolidation may be adopted. Segregated particles appearing on the surface should be removed and used for later construction work. They should not be pushed in the ongoing surface work.

Before concrete hardens, all irregularities are checked and surface corrected by adding or removing concrete. A grout of $1:1\frac{1}{2}$ to $1:2\frac{1}{2}$ of cement mortar is applied on top 20% of pavement to get required finishing.

(*f*) **Curing:** It is important to cure concrete so that concrete develops its strength. The concrete surface is covered with hessein cloth or jute mats and kept fully wet for 24 hours. After that edges are protected with wet clay and bunds are built on the surface with sandy clay to a height of 50 mm. The water is stagnated on the surface and kept wet for 14 days. The traffic is not permitted on concrete surface for another 2 weeks.

(*g*) **Removing of Forms and Joint Sealing:** The forms are removed after curing for 15 days and joints are provided with fillers. Joint fillers may be of soft wood, impregnated fibre board or with cork bound with bitumen. Then a sealer is provided. The sealing compound should be impermeable and at the same time flexible. Bitumen with mineral filler or rubber bitumen compounds are commonly used as sealers.

11.6 BOT PROJECTS FOR HIGHWAYS

BOT means build, operate and transfer. It has been realised that instead of Government itself taking up construction of highways it may be given on BOT contracts to private parties. Government provides the land for constructing the highways, invites the tenders for construction of roads on BOT basis. The total length of road is divided into certain stretches so that private parties can compete for building road. After building the road satisfactorily, the contractor is allowed to collect toll from the users. The rates of toll and period of collections are specified in the tender and finalised in contract. This method of construction has worked out successful since a contractor will take care in all aspects of construction to see that the maintenance cost is minimum. Hence quality road is obtained. After the contract period the road is transferred to the Government and toll collection is stopped. Then onwards, it is the responsibility of the Government to maintain the road.

11.7 ELEMENTS OF TRAFFIC ENGINEERING

Traffic engineering deals with the measures to be taken for safe, rapid and efficient flow of the traffic. The various measures required for it are planning, geometric design, design of crossings, terminals etc. For all these traffic survey should be carried out. In this article, a brief introduction is given to traffic survey, traffic islands and road intersections.

11.7.1 Traffic Survey

This is required to study the type and volume of present traffic and estimate future traffic. It helps in planning expansion of or improvement of the road. The following study is made in traffic survey:

1. Traffic volume
2. Speed
3. Origin and destination of traffic
4. Parking and
5. Accident.

1. **Traffic Volume Study:** It is measured in terms of vehicles crossing a cross-section of road during unit time. The unit of time may be a second to a year. Thus the volume is expressed as vehicles/second, vehicles/hour, vehicles/day and so on.

   However in any design not only number of vehicles but type of vehicles is also important. Hence it is preferred to express volume as equivalent passenger car unit (EPCO) in unit time. The Ministry of Transport has given the following EPCU:

   Car – 1

   Truck, Bus – 3

   Cycles, Motorcycles – 0.5

   Horse-driven vehicle – 4.0

   Bullock cart – 6 to 8.

2. **Speed Study:** Speed is expressed in km/hour. Two types of speed studies are carried out.

   (*a*) Spot speed study (*b*) Speed and delay study.

   (*a*) **Spot Speed Study:** It is required to use in geometric design of roads and also to control speed as per regulation. It is based on measuring the time taken to travel a fixed distance. Enoscopes were used in which stop watch was started when a vehicle starts at specified point and was stopped when it crosses the another specified point. However in this human errors cannot be ruled out. Nowadays electronic meter, graphic recorders and radar speed meters are used which are efficient and indicate speed automatically.

   (*b*) **Speed Delay Study:** This is studied over a longer distance to study the causes of delay in traffic movement over a stretch of road. The delays may be due to singals, road crossings etc. or due to pedastrian crossing, parked vehicles etc. Best method for such study is to run a test car along the traffic recording the duration of delay and reasons for such delay. When a number of such test runs are made real picture for delay emerge.

3. **Origin and Destination of Traffic:** These studies help in planning expansion of roads, bye passes, alternate routes etc. These studies are carried out by interviewing drivers or observing licence number of vehicles or by conducting post card surveys.

4. **Parking Studies:** Parking facility is an important requirement for vehicles. Hence parking studies are carried out to

   (*a*) Obtain information for demand of parking spaces.

   (*b*) For improving existing parking facility.

   (*c*) For planning future parking facilities.

5. **Accident Studies:** Every minor and major accident should be studied for future improvements of roads and signal systems. It helps in deciding making a street one way.

11.7.2 Traffic Islands

Traffic islands are the raised constructions provided to channelise the traffic. They are classified as:

(*i*) Divider islands (*ii*) Channelising islands

(*iii*) Rotary islands (*iv*) Pedestrian loading islands.

(*i*) **Divider Islands:** Figure 11.14 shows a typical divider island which is provided for highways with four or more lanes. They divide the traffics moving in opposite directions and hence prevent head on collisions.

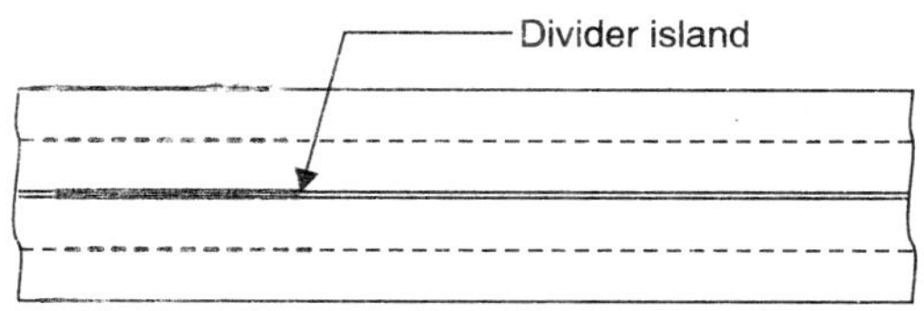

Fig. 11.14 Divider island

(*ii*) **Channelising Islands:** Figure 11.15 shows typical channelising islands provided in the road junctions. They streamline the traffic, provide refuse for pedestrians and serve for locating traffic signals.

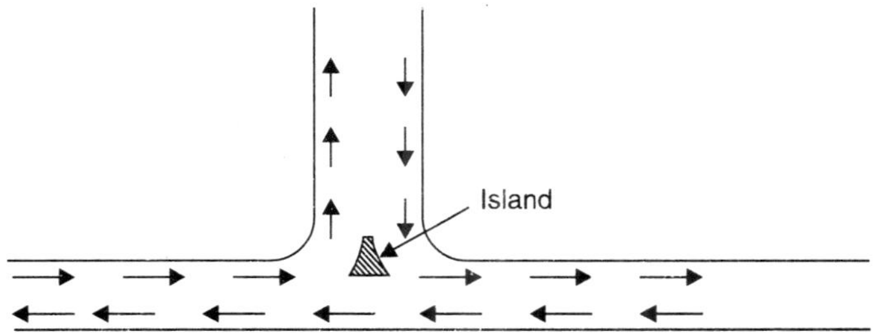

(a)

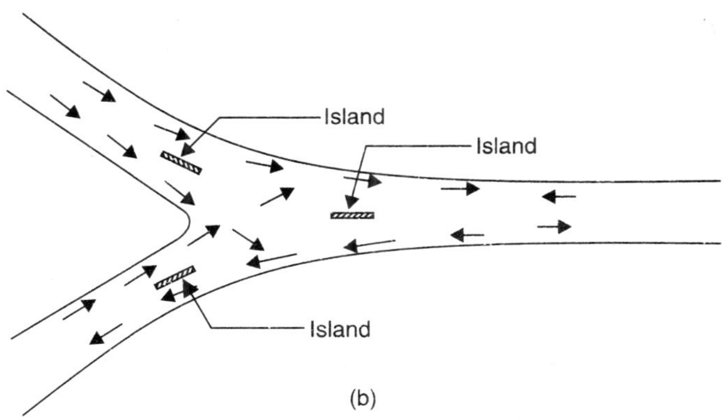

(b)

Fig. 11.15 Channelising island

(*iii*) **Rotary Islands:** Rotary islands are provided in the intersection of roads and directs the vehicles to move in clockwise directions. Thus it ensures smooth flow of traffic. The shape of the central island depends upon the layout of intersecting roads and the available space in the intersection. Figure 11.16 shows a typical rotary island.

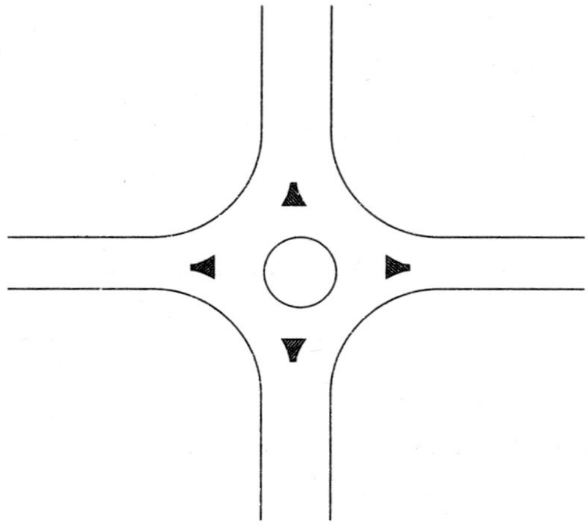

Fig. 11.16 Rotary island

(*iv*) **Pedastrian Loading Island:** These are provided at busstops for the convenience of passengers.

11.7.3 Road Intersections

Road intersections are classified as 'at grade intersections' and 'grade separated intersections'. Road that cross each other at the same level are called 'at grade intersections' while the roads which cross each other at different levels are called grade seperated intersections'.

At Grade Intersections: Several types of at grade intersections are provided. In all the cases it should be ensured that

- Adequate visibility is available
- Proper sign, guard rails and traffic islands are provided.
- Adequate lighting at night is provided.

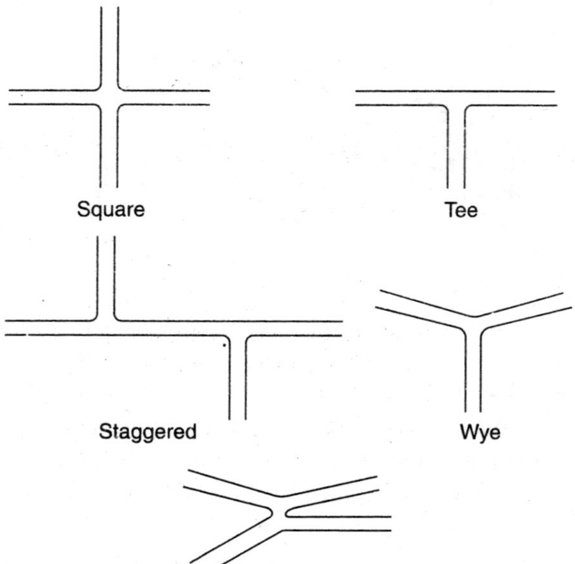

Fig. 11.17 Types at grade intersections

Figure 11.17 shows some of the 'At Grade Intersections'.

Grade Separated Intersections: For important roads grade separated intersections are provided. If major road is above the minor road it is called over crossing; if it is the other way it is called under crossing. The connecting road to these two roads is called flyover. Figure 11.18 shows a typical grade separated intersection with an interchange between the two roads.

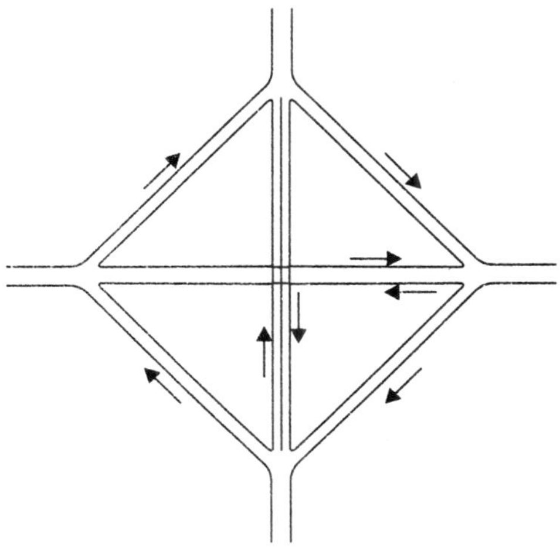

Fig. 11.18 Grade separator intersection

11.8 ELEMENTS OF TRAFFIC CONTROL

Apart from providing traffic islands the following measures are required to regulate traffic.

(*i*) Traffic signs (*ii*) Signals

(*iii*) Markings.

(*i*) **Traffic signs:** Traffic signs are provided in the form of symbols or inscriptions mounted on fixed or portable supports. They guide, warn and guide the road users. The various signs may be classified as

(*a*) Regulatory (*b*) Warning (*c*) Informatory

(*a*) **Regulatory Signs:** These signs are shown on a 600 mm disc installed at a height of 2.8 m above ground level on road side. They may indicate

- No turn
- No entry
- Over taking prohibited
- Sound horn prohibited
- No parking etc.

(*b*) **Warning Signs:** These signs warns the driver and make them to be cautions. They are shown on rectangular boards of size 450 × 400 mm mounted at a height of 2.75 to 2.80 m. Some of such signals show:

- Curve ahead
- Cross roads ahead
- Level crossings
- School zone
- U-turns
- Narrow bridges etc.

(*c*) **Informatory Signs:** These are the sign boards provided to guide drivers to their destination. They may indicate end of speed limit and road name also.

(*ii*) **Traffic Signals:** Traffic signals are provided at intersections in busy roads. They consist of three lights, red indicating stop, amber indicating clear and green indicating 'go' signal. In some places the signals indicate how long the red signal will be there. When the timer comes to zero amber light appears and then within few seconds green light appears. Signals may be operated manually by a policeman or by automatic timing. Traffic actuated signals also may be designed. Pedestrian signals are also necessary at all these signal points to guide pedestrian to cross the roads.

(*iii*) **Road Markings:** Road markings are necessary to guide road users. Indian Road Congress has standardised such markings.

Some of the markings are:

- Centre line of road.
- Traffic lanes
- Edge of roads
- Pedestrian crossings
- Parking space markings
- Bus stop markings
- Kerb markings etc.

QUESTIONS

1. Explain the role of transportation in national development.
2. Write short notes on:
   (*a*) Fair weather road (*b*) Express highways
   (*c*) National highways (*d*) Village roads.
3. Write briefly about the required properties of the following for road works:
   (*a*) Soil (*b*) Aggregates
   (*c*) Bitumen.
4. Briefly explain various steps in constructing a road in embankment.
5. What do you understand by water bound Mechadam Road and Bituminous Mechadam Road ? Briefly explain.

6. Explain the stages involved in the constructions of concrete roads.
7. Write short notes on:

   (*a*) BOT projects for highways (*b*) Traffic survey
8. Explain the following terms with sketches.

   (*a*) Channelising islands (*b*) Rotary islands

   (*c*) Grade separated intersection.
9. Explain different types of traffic signs used.